The Legacy of American Copper Smelting

The Legacy of American Copper Smelting
Industrial Heritage versus Environmental Policy

Bode J. Morin

The University of Tennessee Press

First Edition.

The paper in this book meets the requirements of American National Standards Institute / National Information Standards Organization specification Z39.48-1992 (Permanence of Paper). It contains 30 percent post-consumer waste and is certified by the Forest Stewardship Council.

Library of Congress Cataloging-in-Publication Data

Morin, Bode J.

The legacy of American copper smelting: industrial heritage versus environmental policy / Bode J. Morin. — First edition.

pages cm

Includes bibliographical references and index.

ISBN 978-1-57233-950-7 (hardcover) — ISBN 1-57233-950-0 (hardcover)

1. Copper mines and mining—Environmental aspects—United States.
2. Environmental policy—United States.
3. Copper industry and trade—Tennessee—Ducktown—History.
4. Copper industry and trade—Montana—History.
5. Copper industry and trade—Michigan—Keweenaw Peninusla—History.

I. Title.

TD899.C59M67 2013

338.2'7430973—dc23

2012040202

To my wife, Deborah

Contents

Illustrations

Figures

Tables

Acknowledgments

This project was the culmination of the first Michigan Technological University (MTU) dissertation in industrial heritage and archaeology, a unique multidisciplinary program dedicated to exploring and understanding the social, cultural, and physical remains of industrial societies. MTU and the social sciences department provided a collegial and engaging academic atmosphere and many of those dedicated Yoopers deserve thanks for their help in the dissertation process. Karen Aho and Gina Stevens made the many administrative forays inside and outside of the department seamless and painless. Bruce Seely, as department head for three of my four PhD years, then dean of sciences and arts, thoroughly encouraged and supported the pursuit of this new PhD. Larry Lankton was always open for a discussion about local history and provided prepublication access to his latest book on Michigan mining; Carol McLennan sat on my exam committee and shared, encouraged, and directed my interests in the aftereffects of mining.

Patrick Martin, social sciences department head my last year, secretary of the Society for Industrial Archeology, president of The International Committee for the Conservation of the Industrial Heritage, and member of my dissertation committee, supported, encouraged, and enlivened academic and extra-academic pursuits, and provided important guidance and insights into the field of industrial archeology and heritage. Susan Martin, who sat on both my MTU masters (1995) and PhD committees, encouraged and supported my work at every level and always provided new, meaningful, and stimulating insights into whatever project I had questions or doubts about. Terry Reynolds, as advisor and chair of both my masters and PhD graduate committees, provided meaningful guidance for researching and writing large and small historical works, was always available to discuss research direction, thesis complications, funding options, and writing style—and was always supportive of departmental goals and activities.

It is, of course, difficult to acknowledge everyone else who helped, encouraged, or supported this project, but several key people and institutions deserve recognition and thanks, especially the archives, repositories, and museums I visited. Ken Rush, the Ducktown Basin Museum director, provided invaluable local heritage insights, documents, and photographs of the Tennessee copper industry. Ellen Crain and Lee Whitney of the Butte-Silver Bow Archives provided key direction and access to their Montana collections, important contacts in the state, and important insights into the history and heritage of their community. The Michigan Technological University Archives and Copper Country Historical Collection archivists Erik Nordberg and Julie Blair provided guided access into the MTU collections as well as key insights and engaging discussions about regional mining and tourism history. The several libraries and institutions that maintained regional Superfund public repositories included the Ducktown, Tennessee, City Hall; the Montana Tech Library in Butte, Montana; and the Portage Lake Public Library in Houghton, Michigan. Helpful federal document collections included the EPA Montana Superfund office in Helena, Montana, the EPA Region 8 library in Denver, and the National Park Service Technical Information Center in Denver.

Several individuals also provided meaningful insights and discussions about their work and their respective communities. I would like to thank Sara Sparks of the EPA in Butte, Brian Shovers of the Montana State Library, Lori Hallauer of the EPA office in Helena, Jim Corless, former superintendant of the Keweenaw National Historical Park, Fred Quivik of Michigan Technological University, Paul White of the University of Alaska, Anchorage, and John Sesso and Chuck Carrig of Butte-Silver Bow County for their helpful direction and insights.

I would also like to thank the Americana Foundation for its financial support for the dissertation, the National Science Foundation for financial and travel support, the Michigan Technological University Graduate School for financial support, and the social sciences department for financial, administrative, and travel support. Further, I would like to thank the University of

Tennessee Press for encouraging this publication and the internal editors and outside reviewers who provided meaningful insights into content and format.

Lastly, I'd like to thank my family, especially my parents, Pat and Glen Morin, for their encouragement and support, and the former Deborah Haddrill for her enduring patience, encouragement, and good humor over the years I worked on my dissertation, and the now Deborah Morin for her continued support and good humor as I worked that dissertation into this book.

INTRODUCTION

> Essential for social identity and collective purpose, heritage enriches us through remembered precursors and prospective heirs. But these enduring benefits blind us to a mounting backlash. Age-old aversion toward husbanding the past today grows more virulent. Nature conservation arouses similar hostility, but animus against heritage is harder to counter. Environmentalists can threaten global extinction; heritage advocates warn merely of lower quality of life. To many that seems a lesser, even a negligible threat.[1]
>
> —David Lowenthal, 2000

In its simplest form, this is a heritage study. It examines the debates about which things a community should save, how the final decisions are influenced and negotiated, and what, if anything, should ultimately be remembered about the past. These are not simple or uncontested questions because the significance of past events and historical materials are not universally appreciated. This book uses a specific genre, industrial heritage, and the complicated perceptions of value in an often disheveled and blighted landscape to explore these questions.

Industrial heritage poses challenges that other types of heritage do not. In terms of merely conserving industrial structures, complications of process interpretation and structural deterioration exist in forms and scales that simply do not affect historic house museums or century farms. Further, industrial sites can encompass entire landscapes, which because of increasingly complex and historically significant operations in the late nineteenth and early twentieth centuries, often include some contaminant or hazardous waste threat to human health or the environment. All of these conditions tend to complicate the historic industrial setting, leaving a tangled legacy in need of both commemoration and remediation.

This book will explore how heritage-minded groups living in historic industrial communities and wishing to commemorate their past and preserve their historic resources, articulate and negotiate that desire in the face of sometimes overwhelming opposition. In the cases studied here, protracted opposition will arise from the Environmental Protection Agency (EPA), a well-funded outside federal organization whose mission is to protect human heath and the environment. The EPA, aiming squarely at remediating the waste streams emanating from the landscapes containing those very same historic resources, can at times view heritage preservation as an obstacle to the most efficient remediation—a viewpoint often endorsed by development-oriented community groups.

Peter Howard defines *heritage* broadly, as anything—including material objects, ideas, customs, and practices—that anyone wants to save or remember for any reason.[2] But not all heritage can or should ultimately be preserved especially with the limited funds available to preservationists. To ensure that funds are applied most appropriately, heritage standards and practices dictate that preservation decisions be initially qualified by the relative importance, or cultural significance, of the object, idea, custom, or practice to the community desiring to save it. *The Burra Charter*, drafted by Australia ICOMOS, defines *cultural significance* as the "aesthetic, scientific, social, or spiritual value for past or future generations . . . embodied in the place itself, its fabric, setting, use, associations, meanings, records, related places and related objects."[3] Although these definitions are necessarily broad, for the purpose of this study, *heritage* is simply reflected in the recognition that historical events represented by surviving material culture have helped shape the present, and that those material things and events that have the most significance, relevance, and integrity are worthy of remembrance and some sort of commemoration, documentation, or preservation irrespective of the environmental conditions in which they may reside.

The ultimate decisions about what to commemorate, document, or preserve, however, are neither simple nor straightforward. Where historic evidence defines the age, function, and associations of historic resources, and scientific evidence defines the existence of hazards within a landscape, opinions and values often affect and influence both preservation and remediation decisions. Historians and sociologists have demonstrated that the resolution of technical problems is not a linear process predetermined by its solution but rather a process negotiated and contested by divergent groups without predictable outcomes.[4] This conclusion does not imply that sides are not drawn with some groups strongly favoring one solution over another, but that final work plans are—or can be—significantly influenced through a negotiated political process and that there is never only one course of action to solve a technical problem.

Articulating a clear position, however, can have its challenges. The terms *significance* and *relevance* are subjective in a heritage context, often enumerated by associated values to specific groups at certain times. *Contamination,* too, can be defined by its associated values and perceptions. Although few people would likely argue against the need to assuage the dangers posed by certain carcinogenic toxins, the use of the word *contaminated* to describe lesser or only superficially dangerous landscapes can overstate potential threats to some, whereas to others any contamination is too much contamination. Used in a political context to sway public opinion and perception, arguments for the remediation of contaminated landscapes in the name of human health can be a powerful tool for defending any remediation plan that may or may not include the consideration of historic structures.

Although it is not intended to judge the choices declining communities make, this book will accept that heritage practice and environmental remediation, broadly defined, are good for a community and that the better a community can understand the contamination dangers it faces and define, manage, and protect its heritage, the stronger its postindustrial identity and economic recovery will ultimately be. To understand this decision process and the disputes that arise from it, this book will consider historical events leading to industrial heritage conflicts by briefly examining the broad technological and environmental history of copper smelting, then looking more specifically at important historic copper mining and smelting communities in Michigan, Tennessee, and Montana. Each of these three districts has existing heritage plans that are enriched and contested in various ways by the environmental consequences of their respective mining periods. This book will focus specifically on the issues of smelting technology, waste generation, environmental remediation, and cultural heritage preservation.

Chapter 1, "*The Historic Roots of Copper Production and Smelter Pollution*," addresses basic copper production practices and early smelter pollution issues brought on by the two primary historic copper smelting methods, the Continental or German blast-furnace system and the later Welsh reverberatory furnace system. Chapter 2, "*The American System: a Technological Context*," provides an historic context for understanding the cultural heritage significance of the mining and smelting landscapes that experienced chronic environmental problems discussed in Chapters 3, 4, and 5. This chapter tracks the adaptation of European copper smelting technology in the United States then looks more closely at smelting technologies in the three American copper districts starting with the Michigan Copper Country on the Keweenaw Peninsula in Michigan's Upper Peninsula, which began mining in the 1840s, the Ducktown district in southeast Tennessee, which began mining in the 1850s, and the Butte-Anaconda district in southwest Montana, which began copper mining in the1880s.

The following three chapters look more closely at these regions, exploring their later copper mining and smelting histories, which led to significant waste disposal problems and to the discovery of major environmental problems after each district had considerably slowed metal production and ultimately wound up being listed on—or pursued for inclusion on— the Superfund National Priorities List (NPL). The chapters, organized chronologically based on each district's initial engagement with contamination and remediation of its copper smelting landscapes, also examine early pre-Superfund heritage plans and later heritage-environmental conflicts as communities attempted to preserve elements of their past while remediation professionals worked to clean up the landscape.

Chapter 3, "*Ducktown*," discusses the district's early twentieth-century lawsuits, its 1920s response to seriously degraded landscapes, 1990s efforts of the potentially responsible party to keep the location from being officially labeled "Superfund," and the privatization of both the cleanup and later heritage activities. Chapter 4, "*Montana*," discusses its early twentieth-century lawsuits, the placing of the Butte-Anaconda district on the Superfund National Priorities List (NPL) in 1983, and the heritage plans and preservation projects that have occurred since the 1960s. Chapter 5, "*Quincy Smelter*," discusses the Michigan Superfund project that began with an NPL listing in 1986, but whose last remaining historic smelter was not considered for remediation until 2004, when efforts to include it in the national park fell short and asbestos issues grew more important. Chapter 6, the "*Conclusion*" summarizes the book's main sections and the *appendix* introduces a *Mining District Heritage Model,* a predictive method that articulates the historical processes rural mining and smelting districts undergo as mining peaks, diminishes, and ends—leaving communities with lower population, greater economic distress, contaminated landscapes, and poor external images—and communities attempt to respond to these problems.

Why Copper Smelting?

Ultimately, this book will demonstrate that despite the well-meaning intent of federal laws designed to conserve heritage and protect the environment, not all people share the same values, perceptions, or interpretations of the landscape, and conflicts arise without clear, predictable outcomes. Although conflicts between heritage planning and environmental remediation play out under a variety of circumstances, copper smelting proves an ideal subject, not only because copper was one of the most important and heavily mined metals throughout history and its American sites retain high historical significance and integrity, but also because its extraction and processing have led to some of the worst environmental degradation in this country.

From the start of the metal age through the early Iron Age, copper was the most important metal for many cultures. When combined with tin to produce bronze, the metal was strong, durable, easily cast, and resistant to corrosion. Although significantly less plentiful than iron, which ultimately replaced copper alloys for most tool, weapon, and structural uses by the Roman period, copper's high thermal and electrical conductivity made it very important both for nineteenth-century steam boiler construction and for the rapidly expanding electrical industries in the twentieth century. Copper was responsible for the first large-scale mining boom in the United States, which quickly led to U.S. global production dominance because of the nation's vast reserves, innovative adaptation of European technology, and ever-increasing demand. Further, because of the toxins associated with its ores and processing, the affects of copper production led to very early concerns about landscape degradation across the world, and ultimately to some of the largest Superfund projects in the United States.

The three sites selected for this study represent historically significant copper mining and smelting regions that pursued heritage planning amidst Superfund or Superfund-inspired cleanups. Both the Michigan and Montana copper districts led global production at one time and had—and still have—high historical integrity. Montana's heritage community was and is dominated by local groups, and Michigan's heritage community was formed by a combination of local groups, Michigan Technological University, and the National Park Service. Montana had early sulfur-rich smelter smoke and mill-tailings disposal problems that spread heavy metals and toxins over a wide geographic area, and Michigan, which did not have a significant early pollution problem, ultimately had to deal with its heavy-metals-rich tailings, which were deposited in local waterways.

The Tennessee copper district was for short periods the second or third leading copper producing region in the country and, while it never produced at the rate of either Michigan or Montana, it developed important by-product processes, including globally significant sulfuric acid plants. It, too, has had heritage documentation projects, but its smaller community didn't wield influence over preservation decisions as communities in the other two locations did. Although its contamination problems from smelter smoke and toxin releases into local waterways were nearly as bad as Montana's, the Tennessee copper district's cleanup company negotiated to keep its district from an "official" Superfund designation but still had significant involvement with and monitoring by the EPA and followed all Superfund remediation protocols—except for the cultural heritage considerations mandated for federal agencies.

Regulatory Framework

Heritage protection and environmental remediation are important components of modern cultures, adding critical value to historic places—especially those set within seriously devalued postindustrial landscapes. As such, nations develop policies, enact regulations, and establish practices to promote heritage, remediate contamination, and assign agencies to enforce their laws. The United States and many other nations have heritage laws and professional standards of practice to identify and protect the material things, places, and traditions of the past based on the values for protecting historical structures and landscapes within each country. In most nations, significance is based on these values, and heritage can be significant on a personal, local, regional, national, or global level. Significance and value are thus determined by the meaning a particular thing, place, or tradition has to the culture identifying and protecting it. The home of the first president of the United States, for example, is a key physical representation of the roots of this nation. As such, it has great symbolic value and, hence, great significance on a national level and is protected, preserved, and interpreted in a manner befitting the most important historic structures in this country.

Cultural heritage laws and practices evolved in the United States based on and stemming from traditions established for the preservation of natural landscapes with high scenic beauty. The primary national-level responsibility for identifying and conserving our cultural heritage rests with the National Park Service, an agency whose mission also encompasses the protection of significant natural places. Over a century, the United States developed a system to identify and document structures and landscapes of regional and national significance, and to preserve and interpret significant objects and places held publically and privately. The U.S. Antiquities Act of 1906 first permitted the official designation of historic landmarks and created penalties for harm caused to federally owned cultural resources. Sixty years later, the National Historic Preservation Act of 1966 (NHPA) authorized the formation of the National Register of Historic Places, established rules for managing historic resources on federal property, and created and assigned duties to the State Historic Preservation Offices (SHPO). Further, and perhaps more importantly, the act established regulations directing federal agencies to consider and mitigate the impacts of their actions on cultural resources irrespective of ownership.

Although some states have state-level registers of historic structures and some communities have historical societies, historic districts, preservation boards, and ordinances to protect cultural resources, the most critical list for heritage designation in the United States is the National Register of Historic Places. The register includes sites and structures added to the list after its creation in1966 and National Historic Landmarks designated prior to 1966. To be eligible for listing, a privately or publically owned site must meet certain criteria of regional or national significance, be in its original location, and largely be free from major ornamental or structural changes since its time of greatest importance. Sites on the National Register that meet an even higher level of national significance can be elevated to National Historic Landmark status (while remaining on the register). As of 2011, over 80,000 properties, representing over 1.4 million individual resources, are listed on the National Register, but fewer than 2,500 of them meet the criteria for National Historic Landmark status.[5] Although these designations provide little legal security for privately owned sites—beyond a general protection from federal impacts—a listing at a minimum requires some form of documentation that proves a site's significance and authenticity, thereby fulfilling a key heritage step by creating a record of an important structure or landscape.[6]

The most critical parts of the NHPA for this book are Sections 106 and 110.[7] Section 106 requires that all federal undertakings, defined as projects executed with federal funds or requiring federal permits or licenses, consider and mitigate the potential damage to cultural resources worthy of inclusion in the National Register. Section 110 designates the heads of each federal agency to be responsible for and develop plans and procedures to protect historic resources eligible for the National Register within or affected by their own agencies. The NHPA further established and requires State Historic Preservation Officers (SHPO) in each state to determine the National Register eligibility of potentially impacted cultural resources and to approve mitigation efforts, and provides a National Advisory Council to address national heritage issues, advise the president and Congress on historic preservation, and resolve conflicts. If, for example, the Air Force desires to construct a new hangar building at a base in the United States, and the desired location includes the archaeological remains of an eighteenth-century Native American campsite meeting the National Register eligibility requirements, then the Air Force must consult with the SHPO and try to avoid adverse affects to the site. If adverse affects cannot be avoided, then the Air Force must reach an agreement to mitigate the damage to that site. Mitigation does not necessarily mean cancelling a project, but, based on the site's significance, mitigation could mean, at a minimum, excavation to recover material culture and spatial data prior to construction or, at a maximum, if the site proved to be associated with a significant historical figure for example, finding other ways to execute the hangar project without impacting the site.

Privately owned and nonfederal public sites that are not subject to federal undertakings fall under different consideration. Although no specific federal laws exist to formally protect these heritage sites in this country, tax incentives and certain community grants encourage private historic preservation efforts, and local historic districts can enforce protective covenants or ordinances affecting included properties to ensure that they are protected and maintained appropriately. States, municipalities, and even some private organizations, however, must follow Section 106 if their historic sites or landscapes are developed or modified with federal funds or involvement.

Overall, laws and practices exist to encourage or protect heritage in the United States especially if the federal government is involved, but the relative value of individual heritage sites or traditions to the acting agency is almost inconsequential. The laws require consideration of impacts by federal projects on all heritage determined eligible by the SHPOs under National Register criteria in the belief that the preservation of the historic objects, places, and traditions of the past is in the best interest of the nation.

A second set of United States laws enacted in the latter half of the twentieth century stemmed from a national response to the lasting effects of industrialization and its impacts on human health and the environment. These laws evolved in part from a history of nuisance complaints against polluters dating as far back as seventeenth-century Britain, whose common law legal traditions were emulated in the formation of U.S. law. For example, early nineteenth-century Welsh farmers, whose crops and livestock had died because of smelter smoke from Swansea, were among the first to use nuisance laws to attempt to limit industrial pollution. Early twentieth-century American farmers, state governments, and federal agencies similarly sought injunctions against or damages from large copper companies because of significant smelter pollution and injuries to property. The fact that the majority of the early cases brought by individuals was settled, dismissed, or lost by the plaintiffs, and that the governments' cases largely resulted in negotiated settlements, suggests that economic growth was more important to the nation as a whole than was pollution nuisance, until the mid-twentieth century, when the more serious effects of contamination became much more identifiable and quantifiable.

Although they highlighted the potential impact of industrial activity on human health and the environment, these early cases did little to alter the expansive growth of the U.S. economy and spread of evermore complex industrial activity. Not until the later mid-twentieth century did people in the United States begin to seriously question and protest the lasting effects of this industrial activity on the landscape, especially as environmental disasters, illnesses, and deaths related to contamination were deemed preventable and studies demonstrating the effects of long-term exposure to more invisible toxins increased in number and alarm. Beginning in the 1940s, Congress passed laws regulating the output and transportation of hazardous substances and, in 1970, created the Environmental Protection Agency (EPA), whose mission, simply stated, was (and still is) to protect human health and the environment. The EPA was to execute its mission by establishing pollution standards, monitoring industrial effluent, assessing fines, and prosecuting pollution violators.

The Clean Air Act (1970), which substantially expanded the Air Pollution Control Act of 1955, and the Clean Water Act (1972), which substantially expanded the 1942 Federal Water Pollution Control Act, went a long way toward curbing current and future pollution by setting limits to discharges and establishing penalties for violations. The Resource Conservation and Recovery Act of 1976 (RCRA) established rules for the transportation and disposal of hazardous wastes and provided conservation standards for natural resources. Although all three of these acts had implications for current and future pollution, they did little to remediate the prolonged effects of past pollution and existing hazardous waste deposits. Although the 1977 Surface Mine Control and Reclamation Act collected taxes from active coal mining to be used for abandoned mine reclamation efforts, not until 1980 did the government address a broader spectrum of neglected waste.

In 1980, Congress passed the Comprehensive Environmental Response, Compensation, and Liability Act (CERCLA), more commonly known as "Superfund," to identify existing sites of significant pollution, rank the sites based on potential threats to human health, and plan for and execute the remediation of those sites. Further, if the original polluters, their later corporate successors, or current land-holders were economically viable or legal entities subject to U.S. law, then the EPA could label them a potentially responsible party (PRP) and attempt to force them to execute or, at a minimum, pay for the Superfund cleanup. In those instances where a PRP could not be identified, CERCLA and the 1986 Superfund Amendments and Reauthorization Act (SARA) provided public money, the "*super*-fund," raised from chemical and petroleum industry taxes, to pay for the remediation of significantly polluted sites.

For a site to be remediated under Superfund, studies called Preliminary Assessment/Site Inspection (PA/SI) must first determine the extent of the contamination or confirm findings from existing studies. The PA/SIs establish a threat ranking based on potential public health impacts for individual sites and, if significant enough, lead to listing on the National Priorities List (NPL) that officially defines a location as a Superfund site. The next step, if the site requires a long-term permanent solution, is a Remedial Investigation and Feasibility Study (RI/FS) that the EPA or PRP conducts to characterize the threat and optional treatments. The final choices of remediation options, after considering public comments and opinions, are listed in the Record of Decision (ROD), a legal public document that the EPA and PRP are mandated to follow. Following the publication of the ROD, work plans, safety plans, and public involvement plans are created to execute the ROD, usually completed under the broad heading of Remedial Design and Remedial Action (RD/RA).[8] One key component of the RD/RA process is the Applicable or Relevant and Appropriate Requirements (ARARS), which require the EPA to identify and enforce other federal and state statutes that apply to the clean up and federal undertaking. Often these include RCRA, the Endangered Species Act, and Sections 106 and 110 of the National Historic Preservation Act.

Roots of Heritage Conflicts

Despite the existence of ARARS, some critics have charged the EPA with, at times, directly circumventing the NHPA in the name of protecting human health and the environment when historic resources pose—or appear to pose—an obstacle to the most desired or simplest remediation actions. Others have claimed that the EPA abides by Section 106 only when the effort benefits its mission. Often, these critics and other heritage proponents call for modifications to remediation plans that could succeed in both cleaning up the environment and meeting the Section 106 requirements for considering historic resources. However straightforward the NHPA is, federal agencies are mission driven, shaped by the personalities of their various staffs, and not necessarily inclined to abide by other federal regulations or appreciate local values and heritage as much as the communities they work in do.

Further complicating heritage practice in contaminated communities, the EPA in general does not have a strong institutional understanding of heritage. In 2008 it had a small two-person heritage staff within its national compliance and enforcement division that served the remaining 17,000-person EPA.[9] In almost too simplistic terms, the heritage staff distributed a pamphlet titled, "Historic and Archaeological Resource Protection for USEPA Personnel" with chapters titled, "Why Should you care about Historical and Archaeological Resources?" (Answer: It's the law, regulations require it, it's policy, and it's a good idea), and "What are 'Cultural Resources'?"[10] Few other training resources are available for the EPA or state-level environmental departments to access heritage policy and practice beyond a single preservation remediation course taught by the private National Preservation Institute and a book by Thomas King titled *Cultural Resource Laws and Practice* (2008), outlining regulations including CERCLA that affect cultural resources. King's book, however, is geared primarily toward heritage professionals.[11]

Although the EPA is not well positioned institutionally to respond to heritage issues at Superfund sites, many communities are equally unfamiliar with heritage and environmental laws—or are not positioned well enough to use them to their advantage. Without a strong voice from the community to force EPA compliance with Sections 106 and 110 of NHPA, confrontations become one-sided and heritage is often lost or only minimally considered.

Setting

Some of the 11,000 sites listed on the National Priorities List are military bases, warehouses, or transportation sites, but the majority of them were listed because of long-term industrial activity and contamination. Specifically because of this longevity, many of these places also hold historic significance, as defined by National Register criteria, to American economic, social, technological, and industrial history. Arguably, much of America's importance on the world stage can be attributed to its rapid growth and industrial output between 1850 and 1950. American production affected the outcome of two world wars, changed the structure of global wealth, shifted the seats of world power, firmly established a middle class, and, by and large, created the consumer economy. All of this work took place over the course of one century with little regard for the disposal of the noncommercial by-products of production. Factories dumped wastewater into rivers and oceans, vented acidic gasses into the atmosphere, and allowed chemicals to seep into soils and groundwater. Mines and smelter sites, often in sparsely populated areas, simply left their waste tailings and slag wherever they could and usually only considered them again if they interfered with production or could be recycled for additional profits.

Many historic industrial sites therefore have dichotomous legacies. On the one hand, they hold significance to American history and mark important economic, social, technological, and industrial development and, as such, are significant heritage sites. On the other hand, they are responsible for many of the most egregious pollution releases requiring extensive cleanups that incur considerable cost. Federal laws articulated by the NHPA and CERCLA exist to protect and correct these legacies for the betterment of the public.

To provide a background for understanding how historic copper production created the environmental problems that would eventually lead to remediation efforts, the opening chapter of this book addresses the broad environmental history of copper smelting. Drawing mainly on the work of numerous scholars, this chapter provides a synthetic account of technological developments in the field and their environmental impacts from antiquity to the nineteenth century, leading to a description of American copper smelting practice.

CHAPTER 1

THE HISTORIC ROOTS OF COPPER PRODUCTION AND SMELTER POLLUTION

Copper has been a highly valued commodity in the global economy for eight thousand years. Its use has expanded as its specific properties—ductility, alloy-ability, corrosion-resistance, and conductivity—were successively exploited. Despite its utility, the production of copper consumed significant resources and caused visible and not-so-visible human health threats and environmental degradation from the metal's earliest uses. Although the first smelters were small and their individual impacts minimal, as production increased both in number of furnaces and concentration of smelting works, the deleterious and environmental side effects of production increased as well.

Across the globe, copper is found in three basic compounds, including native, oxide, and sulfide ores, each in combination with beneficial and sometimes detrimental or toxic components that need to be removed during smelting. Native copper is pure and, in general, only requires simple smelting to remove attached waste materials and purify the metal. Oxides are copper-oxygen compounds that are also relatively simple to smelt because oxygen easily combines with carbon in the furnace fuel and is dispelled as carbon dioxide. Sulfides, which are the most plentiful copper ores on earth, are complex copper-sulfur compounds that need to be roasted or burned separately, sometimes for weeks prior to smelting, to drive off the sulfur in combination with oxygen as sulfur dioxide. Other elements often found in copper ores that need to be removed include gold, silver, iron, arsenic, antimony, cobalt, nickel, silicon, and lead.

Copper smelting, the art and science of extracting and refining the metal from its ore is a 5,500-year-old profession. Although still debated, the precise origin of smelting technology was likely ancient Anatolia, which included parts of present-day Turkey and Iran, and spread across Asia, North Africa, and Europe. Early smelting was based on crude short blast furnaces that produced small but easily transported ingots, and became a key trade and military component of the Egyptian and later Roman Empires. These early copper blast furnace technologies ultimately evolved into a standardized, (relatively) high-output, charcoal-fueled, and water-powered industrial practice in Germany during the Middle Ages, and spread to formerly abandoned and newly discovered copper-rich regions across Europe with the diffusion of German metal workers.

This particular technology focused on the blast furnace that smelted copper by combining ores directly with relatively pure charcoal fuel in the vertical furnace stack and withdrawing molten semipure copper and slag waste from the furnace bottom. The long multistage smelting process began with the heap roasting of ores to remove sulfur outside of the furnace, followed by the in-furnace production of several intermediate copper compounds called mattes, each between 50% and 90% pure. The process also included several intermediate roastings outside of the furnace, ultimately resulting in a product called "black" copper, indicating that it came from a blast furnace, which was 95%–99% pure. This copper required a final refining to produce a 99.9% pure copper before it could be used in brass or bronze production (see figure 1.1).

This system, however, could not be successfully implemented in late Renaissance England. Although the country was rich with copper ores, the furnaces' reliance on wood-based, contaminant-free carbon fuels, —something in short supply in seventeenth-century England—severely restricted blast furnace integration. England, however, did ultimately develop a similar system but used a reverberatory furnace for primary smelting and refining. This furnace was more horizontal than the blast furnace and kept the ore separate from cheap, abundant, but impure coal fuel and produced "blister" copper, so-named because of its surface appearance after solidifying. Britain's significant economic growth and industrialization in the eighteenth and nineteenth centuries, its vast empire and rich mineral and coal

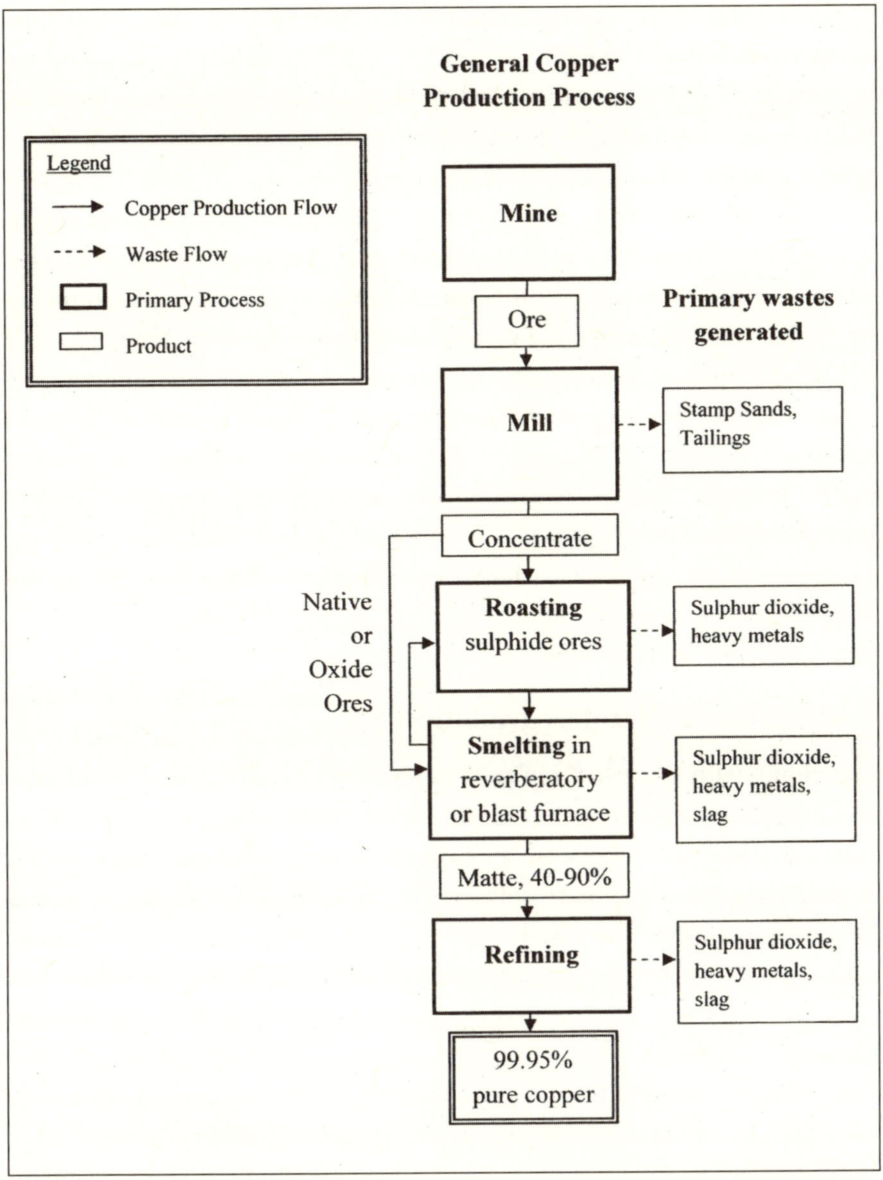

Figure 1.1. Generalized smelting flow diagram. This flow chart for copper production begins with mining ore, then milling to concentrate the ore and remove large portions of waste, followed by roasting of sulfide ores to remove sulfur, smelting to transform the ore into much purer metal, then refining to purify the metal for industrial uses. Although some early processes need to roast and smelt several times before refining, some later processes required multiple refining steps to get the metal ready for electrical uses.

reserves, and a timely slowdown in German production, ultimately led to its world copper domination. By the start of significant American production in the mid-nineteenth century, the two key technological systems based on the blast and reverberatory furnaces were well established and available for further adaptation to suit American needs.

The 5,500-year history of copper smelting has witnessed the application of ever-increasing technological sophistication to increasingly lean ores and the satisfying of demands for ever-purer metal. With such a long history of production and increasingly productive processes, significant changes to the environment were inevitable. Although American environmental remediation efforts at nineteenth- and twentieth-century copper mining and smelting sites, such as Butte, Montana, and Ducktown, Tennessee, are well known in the American environmental community, ancient and historic period writers and contemporary scholars have described the effects of much earlier smelting practices on landscapes from the Roman, German, and Welsh periods of copper production.

Early Environmental Degradation

Coinciding with the growth of trade, increasing populations, and stronger metallic weaponry, people in ancient regions organized politically and began to dominate neighboring regions, forming kingdoms, and then empires. With the growth of extensive political states, the need for metals expanded dramatically, and the expansion of metal production came with ever-greater environmental consequences and landscape degradation. Some of these early problems were obvious to the contemporary populations; others were hidden for millennia.

The most visible landscape change during the ancient period related to deforestation. Mining and smelting activities required considerable quantities of wood for roasting, furnace fuel, and mine timbering. One recent study estimated that, given the amount of slag on Cyprus, metal production was capable of completely deforesting the island sixteen times over a 3,500-year period, although it is very unlikely that the island continuously maintained high production or was ever completely deforested.[1] Based on the volume of slag in Rio Tinto, Spain, another recent study estimated a total timber requirement of 100 million tons for charcoal production alone over the course of ancient and Roman mining.[2]

Another group recently examined the more invisible toxic effluent of mining and smelting activities deposited into the Rio Tinto estuary over five thousand years. This study concluded that elevated levels of copper and zinc existed in water and sediments more than two thousand years before large-scale nineteenth- and twentieth-century open-pit mining. Oysters and clams, they found, survived amid these toxic conditions, providing a direct link between mining contaminants and human intake.[3] Similar studies have been conducted in Massa Marittima, Tuscany, examining unusually high ambient arsenic levels near Roman and later German mining and smelting sites, and in Jordan and Cyprus, tracking metallic bioaccumulation in vegetation from five thousand years of copper and lead production.[4]

The more serious and immediate environmental problem of metal production, however, was not mining but rather the pervasive smoke emanating from smelting works. Early copper and bronze production sites used inefficient open-heap roasting and smelting processes that lost up to 15% of the metal they were producing to smoke alone.[5] For some copper works, the metal tended to vaporize and disperse into the atmosphere with little immediate effect on human heath, but posing more significant long-term soil contamination and bioaccumulation problems.[6] The direct effects of sulfur smoke from open-heap roasting and smelting processes also created critical and immediate concerns for people living near metal works. Although sulfur-specific criticism does not enter the written record until the second millennium CE, early writers did complain about the dangers of mining- and smelting-related air pollution in general. Ancient writers Xenophon and Lucretius observed the noxious emissions from metal mines in Greece, and Pliny declared that smelter emissions were dangerous to animals, especially dogs.[7] Although these examples are not copper specific, all metal working, including precious metal, iron, lead, and copper production, contributed to a Roman edict forbidding metal works in the city because of unhealthy discharges. These conditions ultimately led to the oft-quoted Roman law "Aerem corrumpere non licet," or "polluting air is not allowed."[8]

Falun, Sweden, one of the longest-running mining districts of the last two thousand years—and now a World Heritage Site—first began copper production in the eighth century CE and continued into the late twentieth century. Falun, at its height, operated 140 smelting furnaces and numerous ore roasting stalls based on a system installed and heavily influenced by German smelters.[9] Swedish naturalist Carl von Linne wrote about the area in 1734: "Never has any poet been able to describe Styx, Regnum Subterraneum and Plutonis, nor any theologus hell as gruesome as we can see it here. For outside a poisonous, acrid and sulfurous smoke rises and poisons the air far and wide so that one cannot without pains go there. The smoke corrodes the earth, so that no plants can grow around."[10] Similarly, Brovallius's reported in his 1743 treatise, *Some findings and comments on the smoke from the roasters in Falun,* that the sulfur smell of the roasters could be detected eighty kilometers from the city and that the process completely devegetated the area around the mine.[11]

The high volume and lengthy production in Falun left considerable toxic residue on the landscape. Ek et al. found significant concentrations of mining- and smelting-related copper, lead, and zinc in regional soils especially near mines and smelting sites, thus suggesting significant airborne disbursement. They also found high concentrations of copper, lead, zinc, cadmium, and sulfur in lake sediments, likely from waste dumping or soil leaching. The group estimated that the considerable emissions of sulfur dioxide from roasting and smelting processes over eight hundred years peaked around 1630 and only significantly declined with the advent of sulfur recovery and sulfuric acid production in the nineteenth century.[12]

Massa Marittima, Falun, and several Saxon mine sites all benefited economically from the growth and standardization of German mining and smelting techniques, partially due to similar mining conditions and rich, accessible lodes. In fact, these were some of the longest-running and most productive copper sites in the premodern era. But not all efforts to transplant the German system were successful. Most notably, the attempt to introduce continental-style smelting to England—a region with nearly as long a mining and smelting tradition as Germany—failed due to, among other issues, the difficulty of acquiring a pure source of carbon to fuel blast furnaces amid growing national wood shortages. Ultimately, the British copper industry required the reverberatory furnace, Welsh coal, a concentrated smelting industry in Swansea, Wales, and regional and then globally imported ores to reach world production dominance in the eighteenth and early nineteenth centuries, especially as productive regions using the German system began declining. But, like other areas of long-standing copper production, Wales suffered the environmental consequences of its technological and economic success.

Environmental Effects of Copper Smelting in Swansea

> Delightful Hafod, most serene abode! Thou sweet retreat, fit mansion for a god! Dame Nature, lavish of her gifts we see, And Paradise again restored in thee.[13]
>
> —Anonymous poet, 1737

> Landore. A spot rich in the renown of its metal and chemical works, but to the casual visitor, ugly with all the ugliness of grime, and dust, and mud, and smoke, and indescribable tastes and odours.[14]
>
> —S. C. Ganwell, 1880

The Welsh were not the first to suffer from smelter pollution, but the concentration of works located in southern Wales near very rich coal reserves created critical environmental, legal, and technical problems far greater than any smelter location had previously experienced (see figure 1.2). Although commentators from Roman times through the eighteenth century noted unpleasant odors, dead vegetation, and health problems in areas around smelter works, not until the chronic problems associated with the high output of the Welsh system were copper communities forced to directly address the environmental consequences of smelting. Even Charles Darwin remarked that while visiting Chilean copper mines in the 1830s, "no smoke, furnaces, or great steam-engines, disturb the solitude of the surrounding mountains [as in Britain]."[15]

Smoke was the greatest environmental concern for smelter communities and nearby farmers. In addition to the chemical composition of the smoke, which Piggot described as containing, "sulfurous and sulfuric

Figure 1.2. Swansea, ca. 1860.

acid, arsenious acid, arsenical vapors, fluoride of silicum, fluorine compounds, carbonic acid," the dusts and solid matters contained toxins and heavy metals, such as copper, lead, silver, antimony, and arsenic, that disbursed with the smoke and settled over wide areas.[16] Whereas the gaseous chemicals invaded houses, formed acids after reacting with water in the air, and killed vegetation directly, the solids tended to settle on soils and plant surfaces and were ingested by livestock.[17]

Nineteenth-century health complaints by residents living near Swansea smelters included "a dry sensation in the throat, a bitter, metallic taste in the mouth, loss of appetite, shortness of breath, tightness across the chest, smarting eyes, and frayed tempers."[18] In 1822, doctors coined the term "metal fume disease" to describe influenza-like symptoms found in smelter workers.[19] More recent studies have confirmed the risk of acute respiratory diseases and suggested that increased mortality from nonmalignant respiratory problems existed in populations working in or living near a copper smelter.[20]

Notably, however, reactions to smelter smoke problems were mixed. Some seventeenth- and eighteenth-century British communities initially responded by simply zoning smelters outside of town or into an area where prevailing winds would direct the smoke away from population centers. Some companies, under the threat of litigation, sought technological solutions or bought affected land.[21] Individuals had to resort to the courts and began suing offending firms for *nuisance* violations. The primary nuisance in these cases was "noxious vapors" that caused injury to property—including denuded forests, weakened vegetation, failed crops, and sick or dying livestock—or immediate or long-term injury to human health.[22]

The first English nuisance case was argued in 1608 between neighbors after one created a pigsty on his property. This case led to the landmark ruling that one should "use your own (property) so as not to harm another."[23] Although this case set a precedent that would stand for centuries and influence a rare smelter smoke legal victory when the city of Liverpool forced the Macclesfield Copper Company to relocate its smelter outside of town in 1770, judgments against smelters would generally prove more difficult to win.[24] Only five of twenty recorded smelter pollution cases in Britain resulted in judgments against a smelter company, mostly in the form of modest damage awards.[25] One of the more notable failures was a case brought by a consortium of farmers against Vivian and Son's Hafod works in 1834. The farmers claimed that smoke from the region's largest smelter company destroyed crops, caused health problems, and sickened and killed livestock by causing "Effryddod," or Welsh crippling disease. The farmers ultimately lost the case because they failed to prove that the Hafod smoke was the primary cause of their problems and not, as the defendants' attorneys argued, poor farming.[26] A second suit by the same group a year later resulted in a partial judgment against Pasco Grenfell of the Middle Bank smelter works. The judgment acknowledged some damage from smoke, but, because the plaintiffs could not prove exactly what was damaged specifically by Grenfell's smoke, the judge only awarded the farmers a single shilling.[27]

Confusion about the potential health problems associated with copper smoke tended to complicate lawsuits against smelter firms. In 1842, a royal commission report claimed that smelter smoke inhalation posed no adverse health affects, and an 1845 commission reported that smelter smoke had kept Swansea free from the cholera epidemics in the early 1830s. Morag-Levine suggests that there may be some truth to this latter claim, as the sulfur- and arsenic-laced smoke may have served as a chemical fumigant and disinfectant.[28] In 1928, a questionable British medical report perpetuated the confusion by claiming that smelter smoke was "disinfectant and salutary, although ignorant people have a prejudice against [it]."[29] Further, a bath in the sulfurous "yellow-scum covered" quenching water from the first fusion of the Welsh smelting process was once said to cure mange in dogs.[30]

Some even saw a haunting beauty in the smoke. Dr. Thomas Williams of Swansea, in an 1854 study titled *Report on the Copper-Smoke, its Influence on the Public Health, and the Industrial Diseases of Copper-men,* wrote that smelter stacks "emit gracefully-gyrating, white, smoky, and fleecy columns, which circlingly and wideningly ascend to the upper regions of the atmosphere, there to be lost in the purity of invisible air."[31] John Percy, a metallurgist fully aware of the dangers posed by smelter smoke, wrote in 1861, "A dense cloud of white smoke perpetually hangs over the copper-works of Swansea and the vicinity, and occasionally beautiful effects are produced in the landscape when the rays of the sun fall upon it, especially towards evening."[32]

Despite persistent,—and possibly economically motivated—arguments to the contrary, the public ultimately grew to fully recognize the health threats of smelter smoke. However knowledgeable the populace became, *nuisance* complaints still had to prevail over the perceived economic benefits of smelters to local communities—and the financial gains associated with the copper industry in southern Wales overshadowed potential environmental and health problems. In 1914, *Pyrites,* possibly a company newspaper, wrote that "its sulphurous canopy has transformed verdant hillsides into desert wastes . . . But the mischief done has been more than compensated by the immense trade 'opening' which the copper trade gave to Swansea . . . But for it the towns on the Tawe would have remained just a popular little bathing creek with a little additional identity owing to its porcelain manufacture."[33] Even more compelling are selected verses from *The Song of the Copper Smoke,* published in 1871, which presaged later legal rulings regarding smelter smoke lawsuits in the United States:

> I touched the tall tress with my vapoury hand,
> And their leaves drop off, like courtiers bland . . .
> You may search the vale and mountain high,
> There is not a flower to gladden the eye.

The widow's lone bosom I thrill with joy,
As I fill the hands of her orphan boy,
The miner I help in the sunless cave,
By me rich merchants their fortunes save,
Barristers, bankers, and even clod-hoppers
Would feel very small if they hadn't "some coppers."[34]

Thus, not surprisingly, lawsuits brought by individuals or organized groups of farmers often failed to fully cross the "public nuisance" threshold applied by juries and courts more sympathetic to smelting companies and the economic growth of their communities.[35] Despite the lack of court orders to desist or pay damages, many people, smelter operators included, acknowledged the health hazards posed by copper smoke and hoped for a solution to it. Percy quoted John Henry Vivian of the Hafod works, who stated in the 1820s that the suppression of smelter smoke would be advantageous to the town and neighborhood.[36] Later, Percy not only proclaimed that all the Swansea smelter owners lived outside of town beyond the reaches of the smelter smoke, but also that apologists like Thomas Williams, cited above, who did not accept the smoke as an "unmistakable nuisance [or who] pretend that it is not, must either have a peculiar constitution or lie under some strange delusion."[37]

Knowing of smelter smoke danger, yet facing few government restrictions, smelters ultimately handled the nuisance issue in one of three primary ways before going to court. They either did nothing and waited to be forced into action; settled out of court if a case seemed strong or cheap enough; or, under a more serious threat of lawsuit or—more often—if the prospect of a financial return on by-products existed, applied or attempted a technological solution.[38] The first attempt at by-product recovery from smelter smoke came in the late eighteenth century from a partnership between John Champion and the Macclesfield Copper Company, the same company that lost the first nuisance judgment against smelter smoke. Champion developed a system at a northern Welsh mine site to recover sulfur from initial calcining operations by roasting ore in long kilns, collecting and condensing the smoke, and returning the "cleaned" ore to the mine for sale. While the market for sulfur had yet to fully develop by the eighteenth century and the success of sulfur recovery had not yet been described, the process improved the ore quality, lessened the destruction of local vegetation near smelters, and remained a notable component of the works through the late eighteenth century.[39] Despite this nascent attempt, however, successful and profitable sulfur recovery was still a century away.

The first technological attempt to recover sulfur with any promising potential came from Gossage towers, which in the mid-nineteenth century were successfully condensing and recovering hydrogen chloride from alkali works smoke with a cool water spray. The challenge of adapting this water-spray method to copper smelters, however, was in combining sulfur dioxide, the primary component of smelter smoke, with water.[40] John Henry Vivian of the Hafod works had proposed a similar process in response to an 1821 prize for "obviating the inconvenience arising from the smoke produced by smelting copper ores."[41] Vivian's system used a series of long flues and chambers with water showers to cool the smoke and promote the condensation of sulfur, arsenic, and copper. Although this system was only moderately successful and was not awarded the prize, the long flue concept, generally combined with a single tall stack, garnered favor. In this arrangement, some of the by-product toxins precipitated from the smoke before reaching the top of the stack, affording a partial smoke cleansing and recovery of by-product chemicals. With increased stack heights came increased precipitate, but the taller stacks also dispersed smoke and remaining toxins over a much wider area. Despite a decline in the use of tall stacks in the 1830s because of fears of lawsuits from a wider population distribution, they would eventually become a landscape feature synonymous with copper smelters and, as predicted, provoke lawsuits from a wider population distribution.

Of the available nineteenth-century technologies, new furnace designs ultimately offered the most promising opportunities for solving the problem of smelter smoke in Britain. Smelter operators knew that, under the right conditions, if sulfur dioxides could be oxidized into sulfur trioxides, the new gas would combine much more easily with water to produce sulfuric acid ($SO_2 + O \rightarrow SO_3$ and $SO_3 + H_2O \rightarrow H_2SO_4$). The presence of carbon from the coal smoke, however, prevented the right reactions from occurring inside the furnace, requiring new furnace designs to separate the coal smoke from the copper gases. Although the process succeeded, smoke and gas separation in a reverberatory furnace was difficult and expensive, and most smelter operators never attempted to install separators, whereas those that did ultimately abandoned the practice because of limited returns.[42] In the 1880s, some Swansea smelters attempted to pass an electrical current through the flue to encourage metal precipitation.[43] Although it proved somewhat successful, this technology, in many ways, came too late for the British copper industry, which had begun a sharp decline in the 1880s as major new copper works opened in the United States.

Both the German and Welsh systems for smelting sulfides required extensive procedures and multiple steps, indicating a much greater complexity than the simple descriptions offered here suggest. Often these processes required several alternate roastings and meltings that lasted weeks, with a variety of ores added at critical times to remove specific waste products or to recover valuable by-products, before any actual smelting took place to produce a marketable copper. Despite this complexity, global demand for the red metal continued to grow as new industrial uses were introduced and the industry spread beyond its traditional locales. Because of the use of coal, which was significantly cheaper and more plentiful than charcoal, the Welsh process tended to be the most successfully exported technology to newly developing copper smelting areas around the globe. Although German blast furnaces were simpler to build and operate, they tended to be relegated to slag remelting until technological developments significantly increased their productivity. By the start of large-scale American copper production, coinciding with growing industrial demands, the Welsh system reverberatory furnace had come to dominate copper smelting, and most—but not all—new and successful copper smelting ventures implemented some version of it.

In many ways, the Welsh system developed in Swansea was a transitional industrial one. Operating at the end of five thousand years of traditional production and traditional markets, the Welsh system turned copper into a global commodity before giving way to modern demand and ever-increasing technological sophistication. Its technological systems were new, but its early markets, at least initially, were not until the full advantages of coal and coal-fired steam engines changed the economic landscape of Britain, and electricity heralded a second industrial revolution. On the world stage, the Welsh system only lasted a short time, and only a few key features persisted. Soon its complex processes and small furnaces were gone, especially as new techniques were developed to handle very low-grade ores, first in the United States, then around the world. But elements that persisted included the reverberatory furnace, the tall smoke stack, and significant environmental degradation.

Pollution problems associated with southern Welsh smelting highlighted the potential dangers to human health and property. The idea of environmental degradation became codified in contemporary literature describing the Welsh system and, in many ways, attended Welsh technology as it spread to developing copper regions around the world. Initially, as in Wales, the economic benefits of copper smelters often overshadowed their environmental problems. In the United States, for instance, most early large-scale copper mining and smelting occurred in the remote, sparsely populated, and largely under-regulated West. Attempts to limit pollution did not occur until lawsuits forced the issue, and even then the primary solution was that smelter-operating companies simply purchased the affected land or attempted some solution that had financial returns and continued producing copper as they had been. It would take the 1970 Clean Air Act to formally set legal standards for smelter air emissions, but the act occurred too late to prevent much of the damage. Growing recognition that smoke, slag, and other chemical releases had caused long-term environmental problems forced the United States to rethink its environmental responsability and, under the aegis of the federal Superfund program, to attempt to clean—or at least mitigate—the damage caused by smelting, among other industrial activities.

Chapter 2

The American System: A Technological Context

The seriously degraded American copper mining and smelting landscapes that required extensive EPA involvement in the late twentieth century resulted from the ever-increasing and nationally significant role that technology played both in smelting copper ores and in escalating demands for the metal. Were it not for copper's high conductivity, the rapid spread of electricity across the North American continent, and the continued development of applications that produced light, produced or used electromagnetic power, or enabled long-range telecommunications, there may have never been such a high demand for the red metal, and copper companies would have never needed to develop technological systems to successfully process ores containing as little as 0.5% copper.

Although the three districts discussed in this book generally did not work ores less than 2% rich, many of the systems they implemented made even poorer ores economically viable. But processing such lean ores left considerable amounts of waste and extensive destruction on the landscape, including tailings from the presmelting concentrating of the ores, slag from smelting and refining, gas damage from roasting and smelting, and a host of toxic materials from the ores themselves and from their processing. Each of the districts also operated for more than a century, leaving a considerable environmental legacy with which their postmining communities would have to contend. But the very technologies at the root of such environmental destruction, however, helped propel American technological and economic ascendancy, thus leading to sites and landscapes of significant historic value.

This chapter will explore the historic context of copper smelting in the United States, and how this industry gave rise to three historically important areas, whose contributions to American history rightly deserve commemoration and recognition alongside efforts to remediate their landscapes. Specifically, this chapter will examine the broad historical and technological trends in American smelting, which led to the development of successful systems in the mining districts of Michigan, Tennessee, and Montana, and the evolutionary technological changes that were required to maintain profitability as ore quality diminished. Next, it will look more closely at the three individual regions chronologically and at the specific smelting technologies they used to exploit their particular copper deposits. The chapter will then introduce the environmental histories and consequences of their smelting technologies, which led to direct federal interaction either under Superfund or federally monitored remediation.

Smelting USA

There never was a truly American system of copper smelting. After the decline of British copper production and a very short period of dominance by Chile, the United States became the world's largest copper producer and consumer, and maintained that lead for nearly a century. Two primary factors drove this ascendancy: the discovery of significant American copper deposits and the adoption and modification of existing European technologies to fit the new deposits. The "Americanization" of the smelting industry was further propelled by the persistent need to overcome high labor costs with mechanization and by the general growth of large-scale corporations using available capital to foster technological improvements and to create significant economies of scale.

The scale of these operations became increasingly important as rich copper deposits declined and very large deposits of exceptionally lean ores remained or were discovered. These lean ore bodies often contained as little as 0.5% copper and required ever-greater technological sophistication to eliminate up to 99.5% waste and further reduce labor and processing costs to make the extraction and production of copper from such lean ores economically viable. In fact, complex copper smelting would only succeed when enough capital existed to invest heavily in technologies to adapt those processes to the local geology. Both of these conditions coalesced in the late nineteenth century, as very large American copper deposits opened, demand significantly increased, and, to disseminate information and encourage innovation, a technical community developed from expanding mining colleges, mining journals, and industry-wide conferences. While technological improvements led to ever-increasing production, few engineers at the time considered the lasting environmental effects of successful metallurgical applications on such lean ores. But the elimination of such a high proportion of gangue, or waste rock, on such a massive scale posed not simply a short-term disposal problem for mining companies, but also a significant long-term pollution problem for the United States.

By the mid-nineteenth century, the Welsh system in Britain had evolved into a somewhat inflexible process as the availability of British ore diminished. At the zenith of copper production in Swansea, smelters relied on a complex mix of ores imported from around the world to maximize the productivity of existing equipment, technologies, and locally mined coal, but the Welsh industry ultimately failed to adapt new technology to new ore types as global supplies changed. Several British authors argue that this inflexibility was at the core of British copper smelting's decline. Tylecote writes that Swansea profit margins were so high for the "monopolistic" smelters who set prices for imported ores from underdeveloped nations that they had no incentive to change their system until it was too late, and their own workers were recruited by the very same underdeveloped nations, now seeking to create a rival smelting industry.[1] Newell contends that new technologies, specifically in America, significantly reduced the fuel-to-ore ratio, reversing the economies Swansea gained from its nearby coal fields.[2] Smelting at or near the source of ore would become the dominant regional organization in the industry. While the Swansea industry continued refining some foreign copper matte through the 1960s and became a primary British zinc producer,[3] copper ores in general stayed much closer to mines. Further aggravating Swansea's decline, nineteenth-century British smelting technologies and business models crossed the Atlantic to the United States, where East Coast smelters created a smelting industry and began competing directly with Britain for the then-available ores from South America, Central America, and Cuba.

American dominance in copper smelting began with the opening of Montana copper mines and smelters in the 1880s. The earlier Michigan mines produced relatively pure copper, but, their smelter methods based on Welsh models did not advance the practice mainly because they didn't have to.[4] In fact, Michigan smelters would ultimately adopt new technologies, such as electrolytic refining, only after they had been successfully implemented elsewhere and the economics of local smelting justified it. Tennessee, which also began mining in the mid-nineteenth century, imported smelting technology based primarily on German rather than Welsh models—but, like Michigan, it did so without significantly contributing new methods until the turn of the century when radical changes to smelting practice had begun elsewhere and local environmental and political conditions warranted it. Not until the 1880s, when Montana smelters imported labor-intensive Welsh and German practices and found them costly and impractical for their limited labor pool and ore-type, did smelter practice significantly changed in the United States. Further, the scale of production with traditional-sized furnaces proved too small to meet the demands of a rapidly expanding electrical industry and, as richer ores gave way to leaner ones, improved efficiency levels had to be reached in order to produce copper economically and to meet consumption needs.

The American Century

Although some form of copper smelting had existed in North America since the seventeenth century, the first heavily capitalized smelters emerged in the mid-nineteenth century at the two Revere Copper Company works in the Boston area, the Baltimore and Cuba, and Baltimore Copper companies in Maryland, and then in some capacity in Belleville, New Jersey.[5] These early smelting efforts emulated British business and technological models, but were superseded when significant ore discoveries were made in Michigan and Tennessee in the 1840s, Montana in the 1870s, and Arizona and Utah in the 1890s. From 1883 until the 1980s, the United States led all other nations in copper production and consumption, as the rapid development of electricity and electrical applications around the globe coincided with increased capacity in American copper production. In many ways, copper smelting in the late nineteenth and twentieth centuries, though dominated by American companies, was based on a system of international devices and techniques borrowed from earlier copper and metallurgical applications and adapted to new conditions and new deposits. If the earlier German and Welsh processes were focused on perfecting a single technological system that could accept a variety of ore types, the new methods were much more focused on adapting technology to fit the ore. The American system would be defined by much greater technological flexibility, especially as the quality and quantity of the ore changed.

Driving the "Americanization" of smelting technology was the need both to overcome skilled labor shortages and to economically process the vast fields of lower-grade ores once the smaller high-grade deposits had been exhausted. After the 1880s, the most significant new ore bodies consisted of very large deposits of very low-grade porphyries, often less than 1% copper. Even the rich Montana ores, which in the early 1880s averaged from 25%–60% copper, dwindled to less than 8% by 1910, and the pure Michigan multiton copper masses gave way to 2%–5% conglomerate and amygdaloid lodes. As the richness of mined ore dropped, technological improvements sustained production until the ore bodies became too lean to mine and process profitably.

The demand-side of this evolution included a relatively new technology: electricity generated with and transmitted over copper wires to illuminate cities, communicate across and among continents, and power electric motors. American and global demand for electricity and copper exploded in the late nineteenth century, driving the need to find and exploit greater copper deposits. American demand grew so fast in the nineteenth century that much of the output of the expanding U.S. mines and smelters remained in the country, effectively creating a second global market capable of setting prices separately from the London Metal Exchange, which had been the prime market for copper traded across the world for centuries (see figure 2.1).

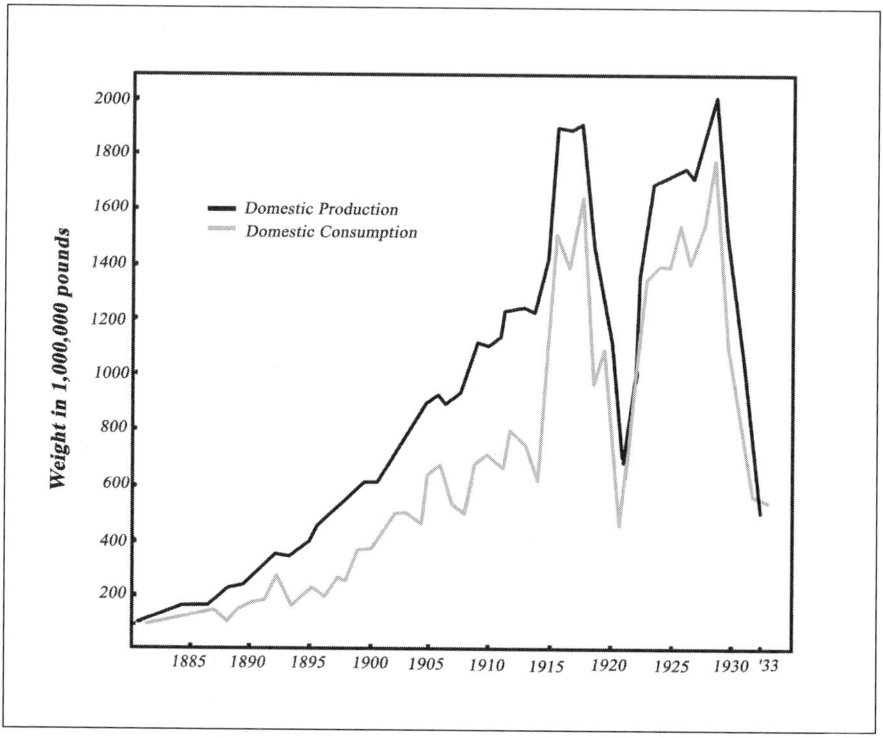

Figure 2.1. United States copper production and consumption, 1880–1933. With the exception of short periods in 1880, 1921, and 1933, the U.S. maintained a substantial trade surplus in copper on the global market. From William Bryan Gates, *Michigan Copper and Boston Dollars,* 1951.

American Smelting Technology

Technology was the key to both increased demand (electricity) and dramatically increased supply. Most early attempts to adapt European smelting technology in the United States drew decidedly on German methods, focusing on the blast, shaft, or cupola furnace[6] (see figure 2.2). The blast furnace held several initial advantages over the reverberatory furnace, including ease of use, relatively low cost, and technological familiarity. The primary disadvantage was that it was limited in size. Through the end of the nineteenth century, the furnaces could not be constructed with a bosh (the furnace hearth where air is added to the heat) wider than thirty-six inches and still maintain the kind of air pressure and temperature required for efficient smelting. This deficiency created bottlenecks in production as greater throughput became necessary to maintain economies of scale. The problem was only corrected by significantly increasing the length of the furnaces—some up to eighty-seven feet—without significantly widening the bosh to maintain adequate width and airflow. The second disadvantage to the blast furnace was that it needed to be charged with larger pieces of high-grade ore to continue sufficient airflow through the furnace. These large ore sizes became increasingly difficult to find as ore quality dropped in mining districts. Lower quality ores required finer crushing to concentrate, usually to a powder consistiency, thus leading to potential airflow problems as the fine ores would clog air passages in the furnace. Despite the increased throughput and efficiency of the larger furnaces, the problem of ore size ultimately led many smelters to start eliminating blast furnaces in the 1920s.

Although the blast furnace may have been the initial choice of most small-scale copper smelters through the mid-nineteenth century, the explosion of American copper demand and production in the following decades

Figure 2.2. Three blast furnace types used in the United States in the nineteenth and early twentieth centuries to process copper. Including (from the left) a typical bottom-dropping cupola furnace, an oval Herreshoff furnace that became the primary pyritic smelter in Tennessee after 1900, and an advertisement for a rectangular furnace that became the standard blast furnace type used in Michigan and Montana where it grew to a size of 87' x 6' by 1920. From Edward Dyer Peters, *Modern Copper Smelting*, 1898.

compounded the need for technological advancement that would significantly increase the throughput of the Welsh style of smelting at a time when blast furnaces faced limited output. Dramatically, a Detroit reverberatory furnace smelting Michigan copper in the 1870s operated at a capacity of 10 tons per day yielding between 4 and 6 tons of copper, and an early 1880s reverberatory furnace in Butte produced 15–18 tons of copper per day. By 1901, the daily output of a single enlarged and improved reverberatory furnace in Montana had reached 100 tons per day, and by 1904, 250 tons per day. Early twentieth-century improvements further increased single-furnace production in Arizona to 600 tons per day. By the 1930s, Arizona reverberatory output had increased to 1,600 tons per day per furnace, and by 1949 Canadian furnaces had reached 2,000 tons per day (see figure 2.3). To remain economically viable, each furnace expansion required increased automa-

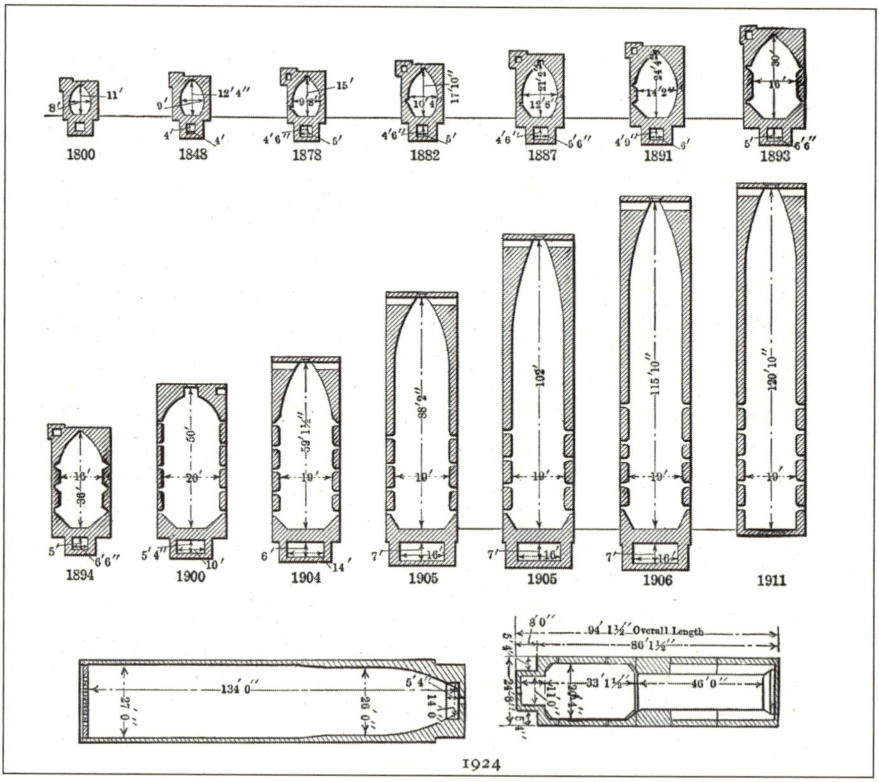

Figure 2.3. The evolution of the reverberatory furnace, 1800–1924. Advances made after 1878 occurred in American mining districts. From H. O. Hofman, *Metallurgy of Copper,* 1924.

tion and handling controls but generated considerable fuel economy with no relative increase in per-furnace labor cost.[7] Even with expanded lengths, individual blast furnaces could not process amounts this large.

Two other technological advancements, flotation and electrolytic refining, would further propel efficiencies in copper production, and each would find its way into the three districts discussed below. As larger pieces of high-grade ore waned, mining lodes often included significant quantities of low-grade ore that required concentration before being charging into a smelter. Flotation, developed in Europe, used very finely crushed ores combined in soapy slurry. The fine high-grade metallic ore particles would cling to the bubbles floating on top of the slurry and could be skimmed off while the waste materials would sink to the tank's bottom. This process resulted in a very high-grade concentrate that would be sent to a roaster or smelter. The electrolytic process, also developed in Europe, involved applying a negative charge to a series of thin pure-copper cathode sheets in an acidic bath that encouraged individual copper molecules to dissolve off of thick, positive-charged matte or blister copper anodes cast from the smelter and set in the same acidic bath. As a result of the applied charge, the cathode, which now included all of the anode copper, was much thicker and, at 99.95% pure, quite sufficient for most electrical applications.[8]

Michigan Smelting with Simplified Welsh Technologies

Historians have argued that processing pure Michigan copper into a molten state should be considered more of a *melting* process than a *smelting* process, whether smelting the large, multiton pieces of pure mass copper from Keweenaw County or the later 2%–5% ores of Houghton County.[9] However, the act of melting pure copper with its attached waste rock in the presence of air tended to change the copper's chemical affinity for other elements, resulting in a copper-oxide-rich compound that required, like all traditionally smelted ores, additional processing to return to a pure state. Ironically, Michigan copper initially became more contaminated because of the smelting process. Following early attempts to adapt Continental-style blast and cupola furnaces with poor results, smelters successfully modified Welsh reverberatory furnaces first to melt large pieces of mass copper, then adapting the systems to handle much smaller copper pieces. After the early richer copper ores diminished further, Michigan smelters began incorporating technological advances developed in the Western United States for lower-grade ores. Smelting the pure native copper of Michigan's Copper Country, however, would remain a fairly simple process throughout the Michigan copper era (see figure 2.4).

Mining companies first attempted smelting in Michigan shortly after the opening of the district in 1844. The region's remoteness and navigational barriers at Sault Ste. Marie created a critical need to establish some sort of localized smelting capacity. Not only was shipping on the upper Great Lakes limited to the ice-free months from April to December, but also the impassable falls at Sault Ste. Marie required all upcoming goods to be offloaded from a lower-lakes vessel, transported above the falls, and loaded onto one of the few upper-lakes vessels, with the process reversed for copper ore and south-bound travelers. The resultant cost of shipping copper ore from the Keweenaw to Boston smelters in the late 1840s ran between eighteen and twenty dollars per ton. The cost to ship one ton of ore from Chile to Boston, however, was only fifteen dollars, and it cost a mere six dollars to ship the same ton from Cuba to Baltimore.[10]

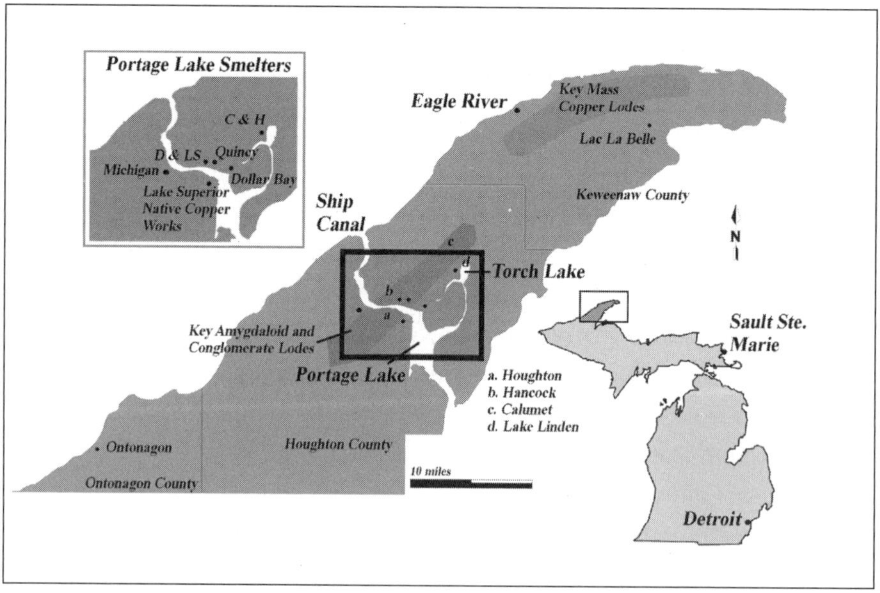

Figure 2.4. Michigan's Keweenaw Peninsula copper country. By the author.

To reduce some of the expense, mines usually built mills to concentrate the rock and remove as much waste material as possible. Although in some cases they were able to achieve fairly high rates of concentration, in general the mines still sent considerable amounts of gangue south. Despite the high cost of transporting copper rock and waste from the Upper Peninsula, only four documented attempts to smelt ore locally occurred before 1860, and each failed largely because their associated mines did not produce enough copper or generate enough capital to perfect the process or justify further investment.[11] Of those four smelters, however, little was recorded beyond rudimentary descriptions of their attempts because neither the smelters nor the mines associated with them developed beyond a season or two.[12] In general, it was still too early for the district, and mines were not producing enough copper ore to warrant the capital expense of building and perfecting a smelter in the region.

Meanwhile, the more successful mines sought an external or contract smelting outlet for their ore. Initially, they shipped large copper masses to the Baltimore and Cuba Company and to Boston's Point Shirley Works. To handle the unique multiton masses produced by the early Michigan mines, the Baltimore smelter installed a side door on one of its reverberatory furnaces; however, the process of charging the large pieces into the furnace damaged the lining so much that the resultant repair costs forced them to withdraw after several trials.[13] The Point Shirley works discovered it needed to cut the large masses into manageable pieces—an expensive and time-consuming task—before charging them into its blast furnace. The extra labor generated a smelting fee of an astonishing eighty dollars per ton.[14] Just a few years later, Detroit smelters would charge less than nineteen dollars per ton.[15]

Prompted by the failure of early Keweenaw smelters and the expense of East Coast smelting, two individuals affiliated with Michigan copper mining successfully developed refining methods a little closer to the Keweenaw. The first was C. G. Hussey, a Pittsburgh doctor and partner in the Pittsburgh and Boston's Cliff, the earliest profitable copper mine in Michigan. After two smelting attempts with little success and hearing reports of a German reverberatory furnace with a removable roof, Hussey hired Welsh smelters to build two reverberatory furnaces with removable rooves to lower mass pieces in for melting (see figure 2.5). Egleston reported in 1881 that the results of the first Pittsburgh casts from the new (1848) furnace were "in every respect as good as those now made" thirty years later.[16]

Curiously, the second part of the Hussey system, a cupola furnace, resulted from a clandestine operation in 1851. One night, according to Egleston, an entrepreneurial local absconded with a boatload of slag from one of Hussey's waste piles. After melting it in an experimental furnace he'd built, he discovered that a significant quantity of oxidized copper remained in the slag and made an offer to purchase all the waste material. Although the person ultimately won a short-term contract to treat the slag in a cupola furnace, Hussey, now aware of the process and its simplicity, erected his own cupola at his site that same year and thereafter treated his slags in house.

Hussey's successful furnace arrangement, including a cupola furnace to treat reverberatory slag, was ultimately copied in Detroit, Cleveland, Baltimore, and a second Pittsburgh location.[17] Not only did the model of a reverberatory furnace for primary smelting and a cupola for slag remelting become the prominent form of processing all Michigan ores for seventy-five years, but also the cupola practice was adopted by the Vivians in Swansea almost a decade after it was implemented in the United States.[18] Although the

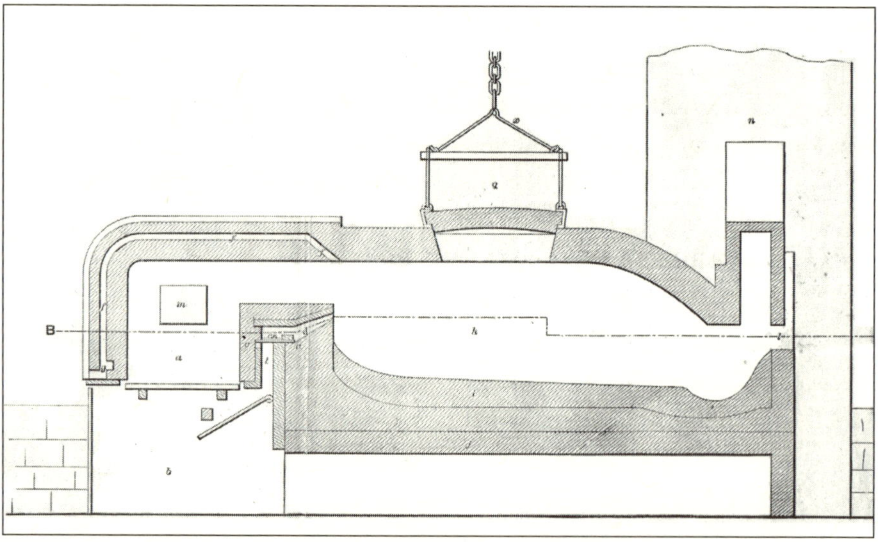

Figure 2.5. Michigan removable-top furnace. From N. H. Egleston, *Transactions of the American Institute of Mining Engineers* 9, 1881.

Cliff Mine and Hussey's other Michigan investments eventually closed, the smelter company he created in Pittsburgh expanded into copper and other metal fabricating and was still operating in 2011.[19]

The second, and ultimately more successful, smelter operation began in Detroit in 1850. Whereas Hussey succeeded in creating a vertical operation, investing in mines, mills, smelters, and eventually rolling and fabrication mills, the Waterbury and Detroit works, under the direction of John Grout, became the leading unaffiliated/contract smelter, processing most of the copper from Michigan mines not owned by Hussey's Pittsburgh and Boston Company. Detroit was located at an ideal location along the primary transportation route from the Keweenaw to the East Coast—in "a position by which all travel and business to and from the mines must of necessity pass."[20] In fact, all the copper rock shipped out of the Keweenaw until 1850 sailed right past Detroit en route to smelters in Baltimore, Boston, or Pittsburgh.

Following several aborted smelting attempts in the 1840s, John Grout eventually opened a Detroit works in 1850 with financial backing from a consortium of Waterbury, Connecticut brass works seeking a consistent supply of high quality copper. The new Detroit smelting works, looking remarkably similar to the Hussey site in Pittsburgh, began smelting purchased and contract Lake Superior copper ores in the fall of 1850.[21] Both works ran removable-roof reverberatory furnaces for the primary melt and cupola furnaces for refining copper-rich slag—and both hired experienced Welsh smelter workers to run their furnaces.[22] Both were also limited to smelting during the ice-free months from April to December.

Three new smelting works on or near the lower Great Lakes opened in the early 1850s, including Grout's Detroit works, a Cleveland site operated by Hussey's brother, and a competing Pittsburgh furnace. All of these enterprises hoped to cash in on Michigan copper production, which increased almost 1,000% from 1850 to 1860 then nearly doubled every decade until the turn of the century. (See table 2.1) However, with the two Hussey sites dedicated to only a couple of mines, and the competing Pittsburgh smelter and the Baltimore and Boston works securing only modest amounts of Michigan ores, the Detroit works held a virtual monopoly on all other Lake Superior copper smelting. During that time, it earned a reputation for innovation and for producing quality copper. Egleston reported that "[a]lmost all the experiments and improvements which have been made in copper refining in the United States since the method was first started in Pittsburg, have been made there."[23] By 1861, the Detroit works had expanded to a capacity of nearly 3,000 tons annually, was capable of smelting roughly half of all Michigan copper, and controlled nearly 40% of all U.S. production.[24]

Through the 1850s, copper mining in Michigan centered on the Keweenaw County mass copper mines with primary shipping out of Eagle River; however, by 1860, copper mining had begun shifting south toward Houghton County. As productivity in the mines producing the large pure mass copper pieces began slowing down, mining efforts significantly increased on the much larger amygdaloid and conglomerate lodes, whose copper contents varied between 2% and 5%, and were found in the areas surrounding Portage Lake, thirty miles south of Eagle River. The companies working the new lodes would ultimately become some of the wealthiest mining corporations of the nineteenth century. Following the opening of the new lodes and a canal and lock system built at Sault Ste. Marie, Michigan production doubled from 1855 to 1860, then doubled again from 1860 to 1870.[25] (See table 2.1)

TABLE 2.1. Michigan Copper Production

Year	Production (tons smelted)	% of US
1850	641	88
1855	2,904	86
1860	6,035	75
1870	12,311	87
1880	24,869	82
1890	50,705	39
1900	72,731	24
1910	110,882	20
1920	80,672	13
1930	84,691	12
1940	45,198	5
1946	21,663	3.6

After William Bryan Gates, *Michigan Copper and Boston Dollars,* 1951, Table 6, 197–98.

Further enhancing production was the region's reputation for high quality "Lake Copper," whose purity for electrical applications set the standard by which all other copper was measured but not able to meet until the adoption of electrolytic refining forty years later.

The success of the Hussey system implemented in Detroit changed the economics of smelting, especially when the system coupled with improved transportation and a significant increase in copper production from the Houghton County mines. There was now enough stable production from the new mines to warrant the capital expense of establishing a local smelter and, with a proven smelting system and much less expensive transportation to import fuel thanks to the opening of the Sault canal and navigation improvements on Keweenaw waterways, the Hancock Portage Lake Smelter opened in 1860 with four reverberatory furnaces and two cupolas.[26] From 1860 until 1867, the Portage Lake Smelter competed directly with the Waterbury and Detroit for all purchased and contract copper ores outside of the mines associated with Hussey.

In 1867, the industry's first significant consolidation took place. The majority of the smelting works outside of Michigan had closed, while Hussey's Pittsburgh works slowed as the output of its associated mines declined. At the same time, the two primary working smelters, the Portage Lake and the Waterbury and Detroit, merged, forming the Detroit and Lake Superior Copper Company (D&LS), which would hold a near monopoly on Michigan copper smelting for the next twenty years. Despite early economic factors that favored Detroit—including cheaper labor, fuel, and transportation costs—the year-round Hancock works that could ship out pure copper—as opposed to ore—expanded rapidly as its operation costs decreased.[27] By 1875, the Hancock smelter operated eight reverberatory furnaces with three cupolas and processed nearly 80% of the company's work, while the Detroit works operated only four reverberatory furnaces and two cupolas and processed the remaining 20% on a continuing seasonal basis.[28]

Early Michigan Smelting Technology

All of the successful smelters processing Michigan copper from 1850 to the 1920s developed systems very similar to Hussey's original arrangement. Copper from the Keweenaw County mass mines shipped as either large individual pieces requiring a removable roof smelter or fist-sized or smaller pieces of mass copper, packed in barrels and appropriately called "barrel work," to Great Lakes smelters. The Houghton County mines shipped some barrel work, too, but the conglomerate and amygdaloid mines shipped their copper ore with its associated gangue as fine, milled, or mechanically crushed ore.[29] The significant impurities included with Michigan copper ores at the time included iron, from wearing milling equipment like stamp shoes, and geologic inclusions of nickel, silver, cobalt, zinc, and lead and eventually arsenic, which existed in veins that cross cut later copper lodes.[30]

Mines usually installed milling or concentrating equipment almost immediately after they began mining in an attempt to reduce the waste material they would need to transport to and charge in smelters. Early Michigan milling consisted primarily of Cornish stamp mills and rollers that crushed and ground the hard rock and some sort of intensive water washing to separate the heavier metal pieces from the lighter gangue. This process resulted in considerable noncopper wastes, called *stamp sands* in Michigan and *tailings* in other parts of the county. In these mills, where the majority of mined copper ores averaged 2–3% copper, the mechanical concentration process could improve the ore's richness up to 75%.[31] The big mines situated stamp mills on waterways for two important reasons. First, they required a significant quantity of water to wash, separate, and transport crushed rock through the mill; and second, the waterways provided a convenient sink to dump the considerable amount of stamp sands generated after keeping on average less than 5% of the materials brought up from the mines.

Through the 1880s and 1890s, following milling, the concentrated rock traveled to one of the contract smelters, but in the twentieth century it was increasingly likely to go to a captive smelter of one of the big mining companies. Initially smelters fired reverberatory furnaces with wood, but by the 1880s, like the Welsh, they used bituminous coal shipped in on vessels that would carry out finished copper. However, unlike the traditional Welsh process, which included multiple reductions to oxidize impurities and fusions to burn off impurities, and refining, the Michigan process only required a single fusion in the melt stage and a single reduction during the refining stage. Most of the conglomerate and amygdaloid smelters were charged with a mix of ore, rich slag between 12% and 30% copper, limestone flux, and black copper (oxide) from the cupola furnace. Slag, mostly removed during the fusion stage, included molten limestone, gangue, portions of the firebrick lining, and impurities that could include nickel, silver, cobalt, zinc, lead, iron, and copper oxide.

After slagging, smelters rabbled the molten bath with twenty-foot long iron paddles. Rabbling, essentially the splashing of molten copper inside the hearth with the furnace doors open, encouraged airflow, oxygen absorption, and the oxidation of any remaining impurities that could then be skimmed off the surface as new slag. Unfortunately, the molten copper also had an affinity for oxygen, and any time the bath was exposed to air, such as during slagging and rabbling, copper oxides tended to form. Smelters often coated the copper's surface with charcoal, a pure carbon, to breakdown existing copper oxides and to prevent the formation of new copper oxides by encouraging the formation of carbon dioxide.

The final stage, reduction, involved charging the copper bath with a fifteen- to twenty-foot green hardwood pole. The rapid combustion of the pole with the furnace doors closed pulled the remaining oxygen out of any copper oxides in the bath, leaving behind a pure metal and vented carbon dioxide. Smelters then cast the copper into either ingots for remelting and/or brass/alloy production, large cakes for rolling into sheets, or rectangular bars for wire and rod manufacturing. The rich slag removed from the reverberatory furnace was charged into a cupola, which by the 1880s was fired with coke and smelted. Workers tapped the cupola slag and discarded it, while the copper that had converted to a black oxide in the furnace was tapped, cooled, and recharged into a later reverberatory heat[32] (see figure 2.6).

Captive Smelting

Although the general technological system for smelting Michigan coppers would remain largely unchanged through the twentieth century, except for expanded-capacity furnaces, advanced coal use, and the addition of electrolytic refining, the economic system of smelting purchased ores or contract ores from mining companies began to change in 1886. Out west, Montana mining firms were erecting their own smelters very near mine sites, and the iron and steel, railroad, and oil industries were vertically integrating by acquiring all of the components of production. While some contract smelting would continue on the Keweenaw after the turn of the century, the general trend was for mining companies to establish their own

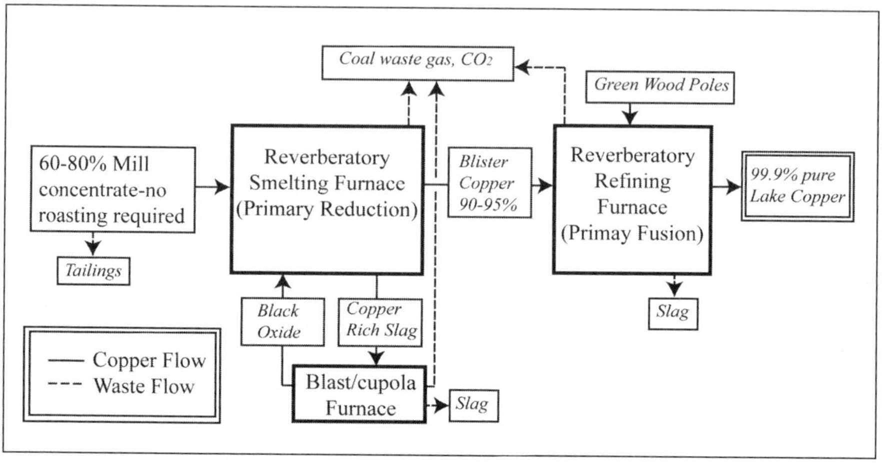

Figure 2.6. Native copper smelting practice on the Keweenaw Peninsula, ca. 1910, based on descriptions in Egleston, *Transactions of the American Institute of Mining Engineers,* 1881.

smelting works and to control the entire process—from mining and milling to smelting and refining.

The Calumet and Hecla (C&H), which began mining as two separate companies on the richest lode in the region in 1865, established the first captive smelter in the area at Hubbell on Torch Lake in 1886 (see figure 2.4). Constructed initially as a joint venture with the Detroit and Lake Superior (D&LS) smelter, the additional capacity of the new regional facility forced the closure of the D&LS Detroit works in 1887. In 1892, C&H bought out the remaining joint venture shares, taking full control of the Hubbell smelter. With five years smelting experience by 1892, C&H opened a second works in Buffalo, New York, to take advantage of lower regional fuel costs and greater electrical capacity.[33]

In 1894, C&H constructed an electrolytic plant in Buffalo, taking advantage of cheaper Niagara electricity to reprocess rich smelter slags and recover silver from the ore.[34] Although the English and German developers had attempted to keep their processes secret, electrolytic refining was adopted in the U.S. by 1883, and three successful electrolytic copper plants operated in Baltimore, and Great Falls and Anaconda, Montana by 1894.[35] In 1914, C&H closed its Buffalo smelter plant and moved its equipment and electrolytic works to its Hubbell location. For many copper smelters, especially ones already working pure copper, silver extraction often justified the expense of constructing and operating expensive plants or equipment like an electrolytic facility. However, when the silver content of its ores declined and lowered the returns from its plant, C&H closed its Hubbell site in 1923 and ended the only Michigan copper electrolysis works.

Opening the Dollar Bay works in1889, the Tamarack and Osceola Copper companies established the second captive smelter in the region. This small smelter merged with the remaining Detroit and Lake Superior Hancock smelter in 1891, forming the Lake Superior Smelting Company (LSSC), which was controlled by the Tamarack, Osceola Consolidated, Isle Royal, and Ahmeek mining companies. LSSC operated its Hancock and Dollar Bay smelters full time through 1903, at which time the Copper Range Consolidated Copper Company opened its Michigan smelter. After five years of intermittent operations, the LSSC closed its Hancock smelting location in 1908. Ultimately, C&H took control of the Osceola Consolidated and Ahmeek mines and, in 1907, became the majority owner in the Lake Superior Smelting Company. The mining company used the Dollar Bay works to smelt the copper of its subsidiary mining interests through 1919 and to experiment with new techniques emerging from the West.[36]

The Quincy Mining Company (QMC) formed in 1846 and prospered from 1856, when it discovered the Pewabic lode on its property. It paid dividends to stockholders from 1862–1864, and then every year between 1867 and 1920.[37] Until 1898, Quincy contracted with the Waterbury and Detroit—or one of its subsequent incarnations—for all of its mineral refining. In 1898, after years of discussion, the firm constructed a smelter on land formed by the waste tailings of the Pewabic and recently burned Franklin mills.[38] Initially, the smelter included three reverberatory furnaces, a cupola building, and ancillary structures, but shortly thereafter Quincy added a fourth reverberatory furnace. By 1904, the company added a fifth reverberatory furnace with greater capacity for low-grade ore in a new building attached to the primary reverberatory house. The smelter works was so successful for the mining company that between 1897 and 1901, Quincy's per-ton smelting costs dropped from $15.30 to $10.06, resulting in a fifty thousand dollar savings in just two years of operation, roughly equivalent to one-third of the smelter construction costs.[39]

Quincy Mining Company production peaked between 1909 and 1911 and, despite high profits during WWI, the firm never fully recovered. Through the postwar economic decline and the 1920s, the firm lost money, eventually closing its mines and smelter operations in 1931, during the Great Depression. By 1937, the firm had begun mining and processing copper again and, with government-set prices through WWII, returned to profitability. Revenues were bolstered with reclamation efforts, modeled after a C&H-developed process that recovered copper wasted from earlier, inefficient stamp milling practices. The system used a dredge to retrieve old stamp sands dumped in the lake and new mills to finely crush the sands, which were sent to a flotation or leaching concentration process to extract the metal. Despite an end to Quincy mining in 1945, the smelter continued to process reclaimed copper until the company's stamp sands were exhausted in 1967, and its contract work, which began after the Michigan Smelter closed in 1948, ended. In 1968, the large number 5 furnace was fitted with natural gas burners and used to melt scrap copper until 1971 when the Quincy chose to shut down rather than abide by new state environmental directives to install pollution controls.[40]

In 1903, the Copper Range mining company, with a growing concern about emission controls, hired a Montana firm with former Michigander Frank Klepetko consulting. Klepetko, who had just finished building the Washoe smelter in Anaconda, Montana, designed the advanced new Michigan smelter with a single flue system to vent smoke and gases from all of its furnaces. While the Quincy smelter largely developed out of a fifty-year smelting tradition on the Keweenaw, the Copper Range adopted a much more contemporary and stream-lined practice and hired outside designers to implement its works even though the technology behind the actual smelting remained fairly simple. The designers, following a pattern very similar to Montana smelters, arranged buildings on a hillside to take advantage of gravity and to eliminate as much labor as possible, and placed a single stack at the top of the hill to consolidate smoke. The tall stack vented smoke higher into the atmosphere, presumably away from air currents that led into populated areas, and allowed for a single dust collection system. The routing of hot gases away from the furnaces also permitted the construction of waste-heat boilers to recover some lost heat, which was used to create steam to run electrical generators.[41] This latter practice was also adopted at Quincy and C&H. Copper Range operated the Michigan Smelter until the postwar economic decline forced its closure in 1948. Although its attention began to turn to a copper sulfide ore lode in Ontonagon County, the mining company continued to mine native copper until 1967, and contracted with Quincy for all of its smelting after 1948.

The formation of the Lake Superior Smelting Company in 1891 largely ended the era of independent contract smelting in Michigan. Although the Hancock works continued to smelt copper under contract through most of the 1890s, and the privately held Lake Superior Native Copper Works smelted smaller mine batches intermittently up through some point in the late 1890s,[42] the construction of the C&H Buffalo works in 1892, the Quincy Smelter in 1898, and the Michigan Smelter in 1903, and the acquisition and later legal consolidation of many of the smaller mines into either the Calumet and Hecla, Quincy, or Copper Range Consolidated Copper companies effectively dealt a death blow to the independent contract smelters. The three consolidated companies came to dominate particular microregions of the mining district, with each relying on a single regional smelting operation to process its entire copper mineral after C&H closed its Buffalo works in 1914 and its Dollar Bay works in 1919. Although a few smaller smelters continued past 1914, by the end of the decade, the mines controlling them would be absorbed by one of the big three and their respective smelting operations would be shut down.

The Michigan copper mining and smelting regions began to decline as the mines' productivity declined, especially as newer approaches to mining very large lodes of especially lean porphyry ores increased in the West. This economic and industrial decline, however, was not replaced by a similarly sized industry and, thus, the population and demand for land dropped significantly. These conditions, while at times leading to difficult prospects for the people who remained, also reduced the need for and financial capability to remove the structures and places from the landscape. Out of these former industrial places ultimately came a movement to preserve them. Because a significant portion of the Quincy Smelter site remained intact and had high integrity and historical value, it became a key contributing site of the Keweenaw National Historical Park, discussed in Chapter 5. Considerably less intact remains were also visible in 2009 at the C&H Smelter, the Michigan Smelter, and the Lake Superior Hancock and Dollar Bay smelter sites.

A Word about Stamp Sands

The waste streams from Michigan copper, as with all other significant copper districts, were immense; however, its smelters did not produce excessive sulfur smoke because its ores were relatively clean and did not require extensive chemical processing to recover its metal. Although no known early lawsuits were filed for property damage—largely because the region was uninhabited by Euro-Americans prior to the mining period—mill-tailings or stamp sands would ultimately prove to be a major problem. Ores with copper contents averaging 2%–3% during the peak periods of production meant that miners hauled 97%–98% waste from below ground. Milling enriched the 2%–3% copper rock to 60%–80% copper concentrate, with copper companies dumping about 95% of mine haulings directly into a lake. Later, they recovered it, reclaimed any copper remaining in the tailings, and dumped the now finer sands back into the lake a second time. In addition to the stamp sands, 2% of what was originally mined was eventually removed during smelting as slag and left on the landscape in similar repositories.

In brief, for every ton of nonmass native copper produced in Michigan, thirty-two tons of tailings and two-thirds of a ton of slag on average were left in the local environment. In 1916, the single year of greatest production, Michigan's copper country produced 13,342 tons of copper, 427,000 tons of tailings, and 8,900 tons of slag. For the period from 1845–1946, the entire Michigan District produced 9.65 million tons of copper, 309 million tons of tailings, and 6.43 million tons of slag, based on an average copper content of 3%.[43] A small portion of the total slag, however, resided near smelters in Boston, Baltimore, Pittsburgh, Cleveland, and Detroit. These tailings numbers do not account for failed mining ventures, of which there were many. It was ultimately the heavy metals in the vast tailings deposits throughout the Michigan Copper Country that raised concerns about environmental toxicity and led to state studies of the region and, ultimately, to a U.S. Superfund designation.

Copper Basin, Tennessee

Unlike the rather continuous development in Michigan copper, external political conflicts and slack infrastructure developments marred and disrupted production in isolated southeast Tennessee through the late-nineteenth century, breaking its mining and smelting history into three distinct phases. The Civil War ended the first phase of regional mining and smelting, and postwar transportation problems ended the second—before railroads finally opened the region to large-scale development in the third phase—and by-product production propelled the district's growth in the twentieth century.

Phase I and II Mining and Smelting

Following the forced removal of the Cherokee Nation in the 1830s, partially initiated by the Band's refusal to concede its gold-rich lands to the U.S. government, the north Georgia, east Tennessee, and western North Carolina region opened for Euro-American speculation and land development. In 1843, prospectors broadened their mineral search north of the recent gold rush in Dahlonega, Georgia and discovered cuprite crystals in southeast Tennessee, which they initially mistook for gold.[44] Although gold was never found in the area, copper was—and within four years, a German speculator made the first rich mineral shipments out of the region, sending his copper ores to the Point Shirley Works in Boston for smelting.[45]

Appalachian copper deposits stretched from Georgia to Vermont but were most prominent and profitable across seven accessible veins in the southeast Tennessee *Ducktown* district, also known by the more geologically specific term, *Copper Basin*[46] (see figure 2.7). The Ducktown deposits exhibited a traditional sulfide profile with an iron-rich cap, or "gossan layer" covering an enriched oxide layer up to 60% copper, over a secondary enriched 10%–20% oxide layer, all sitting above massive 2%–3% iron pyrite (FeS_2) and chalcopyrite ($CuFeS_2$) sulfide layers. Early miners shipped the rich oxide ores to Swansea and Boston smelters profitably, but the secondary layers required milling and an initial local smelting to an intermediate 40%–50% copper matte to make the trip out of the district profitable. The massive 2%–3% copper sulfides would need to be fully smelted and refined locally.[47]

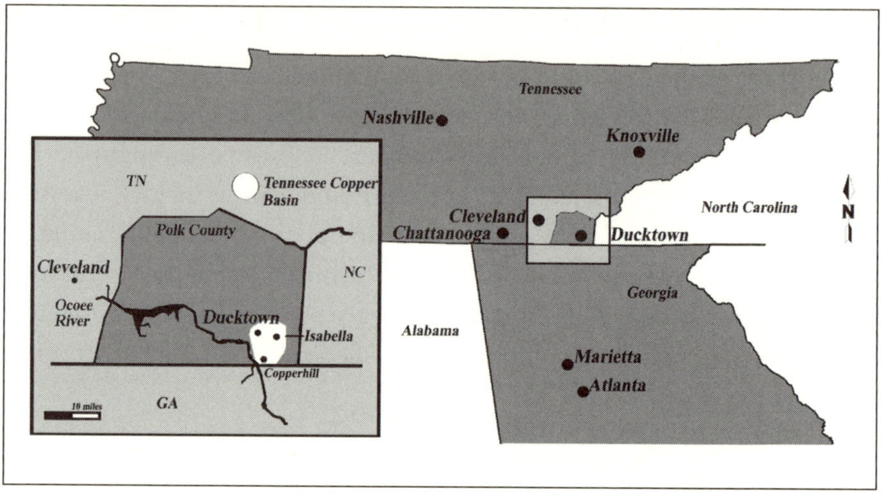

Figure 2.7. Tennessee Copper Basin. By the author.

The remote district became considerably more accessible in 1850, when a railroad reached Cleveland, Tennessee, thirty-three miles west, and investors funded the Ocoee Toll and Plank Road, also known as the *Copper Road,* from Ducktown to Cleveland in 1853. In 1850, the Hiwassee Company officially started mining and, by 1852, three other mines operated in the district. By 1854, following the opening of the plank road, the number of operating mines expanded dramatically and reached fourteen.[48] Despite the arduous and expensive two-day ox-cart journey to transport copper ores and mattes out of the district, a regional industry developed and, by 1860, seven smelters operated in the basin.[49] The few works to discuss Ducktown smelting in the 1850s reference primarily charcoal-fired, German-based blast furnaces augmented with reverberatory calcining furnaces and occasional "Welsh or Swansea pattern" smelters.[50]

Following the economic panic of 1857, the fourteen active mines consolidated into two primary corporations, and, as economic conditions improved, a rolling mill opened in Cleveland, Tennessee in 1860 to produce more finished copper materials and products. Despite positive industrial developments in the 1850s and early 1860s, mining in the district could not survive the loss of personnel and operating capital during the Civil War as young Southerners joined the Confederacy, Northern workers returned home, and investors withdrew their investments for the duration. From 1861–1863, the Confederacy bought all of the copper that the few open mines could produce—until the Union Army destroyed the Cleveland Rolling Mill, which effectively ended copper production in the district and brought the first phase of mining to a close.[51]

In 1866, the two consolidated, prewar companies were reorganized and rechartered. Both the Burra Burra Company, named for the recently discovered Australian copper district, and the Union Consolidated Mine opened with German miner J. E. Raht at the helm. Raht also supervised the only remaining independent mine, the Polk County, for a few years after the war before it closed and hired his brother August to run the smelting complexes at Isabella, and the Tennessee and Polk County mines.[52] At some point in the late 1860s, however, smelting at the Tennessee and Polk slowed down markedly, and J. E. Raht made a significant commitment to Isabella as the primary smelting operation for the district. Not only did Raht build a new refinery at the site in 1871, but he also funded experiments based on an 1866 patent taken out by his other brother William for a Bessemer-style copper matte refining process.[53]

The dwindling ore quality that all mining districts eventually face, climbing fuel prices, expensive ox-cart and plank road transportation into and out of the district, and the high and still increasing costs of operating a copper company in remote southeast Tennessee ultimately forced the Tennessee and Polk County Mines to close by 1868 and the Burra Burra Company to close in 1872.[54] This closure left the Union Consolidated in control of the entire output of copper and the only smelter and refining complex at Isabella. Pressing on under Raht, the company invested in mining equipment and smelting upgrades, ultimately settling on a standard of practice closely resembling the German system with American-modified Welsh refining leading to its most productive year, in 1878.

Phase I and II Smelting Technology

Like all sulfide smelting systems, the Tennessee process used an initial roast to release sulfur. This procedure occurred—as it had for centuries in Europe—with ore piled on top of an open cordwood bed that, once lit, according to one source, would produce up to one million pounds of yellow gas per roast.[55] The first reduction to produce intermediate copper matte took place in "small square blast furnaces" with two tuyeres, or air nozzles, entering the back wall. A second roast was carried out up to five times to further eliminate sulfur, a step that was followed by a second reduction, this time for black copper that was nominally 90% pure. The final reduction occurred in a reverberatory furnace that produced a blister copper (99%), followed by a refining stage in a second reverberatory furnace to further purify the copper for most industrial uses. All Union Consolidated furnaces used charcoal fuel, but due to increasing fuel shortages, by 1876 both blister and refining reverberatory furnaces were fired with gas. Raht also implemented a compressed air system to blast air into the molten refinery bath to oxidize impurities, emulating the Bessemer process patented by his brother and further reducing fuel needs[56] (see figure 2.8).

One reason for the partial conversion to gas was to reduce the enormous quantity of wood being consumed for mining, roasting, and smelting in southeast Tennessee and to better manage the works in the face of its inevitably dwindling supply. An 1872 visitor to the region reported in the *Cherokee Scout* (NC) that the fuel consumed by the Isabella smelter reached sixty thousand bushels of charcoal per month.[57] According to a Department of Agriculture report in 1875, Tennessee forests averaged between thirty and seventy cords per acre, although some exceptionally dense old-growth forests could generate as many as 700.[58] Borrowing a general charcoal production rate estimated by Gordon and Malone at fifty bushels per cord, a conservative average charcoal consumption rate of forty-five thousand bushels per smelter-month, and a typical forest yield of fifty cords per acre, the monthly wood consumption per smelter furnace roughly equaled eighteen acres of forest land.[59] Further, estimating the number of smelters running, the district had operated 756 smelter-months by 1878.[60] At eighteen acres required per smelter-month, smelting likely consumed a total of 13,600 forested acres or twenty-one square miles by 1878. With the basin's size of thirty-six square miles, not all of which was forested, the region's smelters consumed a significant quantity of available wood with no indication of forest conservation planning. With open-heap roasting, mine timbering, building construction, plank-road construction and repair, home heating, and boiler operation, all wood uses led to severe regional timber shortages. Other consumption estimates suggested that mining in total stripped between forty-seven and

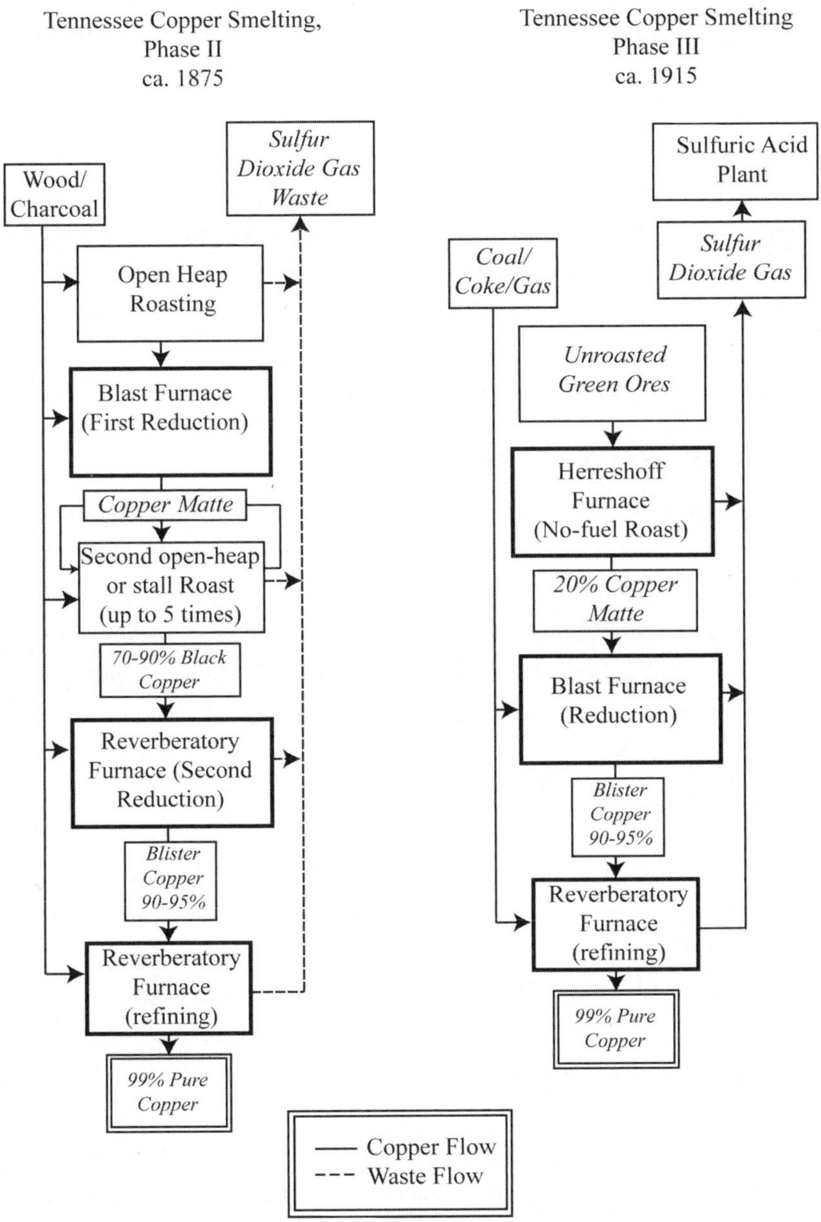

Figure 2.8. Tennessee copper smelting practice. Based on descriptions found in Robert Edward Barclay, *Ducktown Back in Raht's Time,* 1946.

fifty square miles of regional forests for fuel and other uses by 1879.[61] Exacerbating the growing fuel consumption problem, toxins from ore roasting and smelting spread across the landscape, retarding regrowth and new vegetation in the district.

Even as the Union Consolidated would have its best production year in 1878, it too could not sustain operations with the increasing wood-fuel shortages. Despite attempts to lower costs using alternative fuels and imported wood from Georgia, the Union Consolidated's dwindling profitably was further worsened by Michigan's postwar production heights and a greater than 50% drop in the price of copper since 1872. These conditions ultimately forced the last Tennessee copper mine to close in 1878, ending for the second time all copper production in the state.[62]

Phase III Mining and Smelting

> When we consider what may yet be peering into the possibilities of the future when the locomotive shall appear in their midst, bringing in its trains of coal from Chattanooga or Knoxville, and whilst they are amazed at the timidness of capital which fails to see a golden opportunity, they are yet buoyed up with the hope that ere long their best hopes in this direction will be fully realized. Without a railroad we fear that Ichabod is written upon their unequalled and inexhaustible mineral deposits.[63]

This quotation from an 1872 *Cherokee Scout* article on Ducktown appeared the same year that the Burra Burra mine closed, four years after the Polk County Mine closed, and yet still six years before the Union Consolidated would close and end copper mining in the district once again. Foretelling the certain demise of phase II mining six years before its end, the author also recognized the critical production problem for the region—and its ultimate solution.

In 1890, after more than a decade without any copper production in Tennessee, the north-bound Marietta and North Georgia Railroad constructed a rail line to the Copper Basin, which met up with the new Knoxville Southern Railroad south-bound line. With copper demand on the rise since 1885,[64] prices on the rise since 1887, and coal fuel and inexpensive transportation now available via rail, copper mining and smelting returned to the district with the incorporation of the Ducktown Sulpher, Copper, and Iron Company (DSCI) in London, England, in 1891, and the Tennessee Copper Company (TCC) in New York in 1899.[65] DSCI began with the holdings of the Union Consolidated mining company, which included the Mary Mine and Isabella mine and smelter complex, and TCC bought the former Burra Burra, Polk County, and other independent mine holdings.

Shortly after reopening the Mary Mine and the Isabella Smelter, DSCI installed an efficient high throughput 100-ton/day coke-fired Herreshoff blast furnace[66] (see figure 2.3). In 1893, it installed a second Herreshoff and, by 1901, operated four furnaces and was producing three million pounds of copper annually.[67] Despite new fuels and equipment, DSCI continued the old practice of open-heap roasting, which ultimately posed several key problems. First, heap roasting methods required several weeks to process an amount of ore that the new blast furnaces smelted in only a few days, creating a significant bottleneck in production.[68] Second, the sulfuric waste gases from the process quickly dispersed in the basin, creating acidic gases and acid rain, killing any remaining vegetation near the works, and deeply affecting nearby farms and forests in much the same way that Swansea smelters did earlier in the century. The Tennessee situation, however, was exacerbated by several prior decades of rampant timber harvesting, leading to a very distinctive lunar-desert landscape (see figures 2.9–2.10).

Like DSCI, TCC started operations with mining, smelting, and open-heap roasting at the McPherson and Polk County mines near Ducktown in 1901 and, by 1903, operated three blast furnaces and more than 150 roasting sheds. By 1906, TCC would consolidate its smelting operations at Copperhill (likely in 1904), operate a five-mile railroad between its smelter and Ducktown mines, and smelt in seven blast furnaces.[69] Continued open-heap roasting, however, now significantly increased by TCC, proved to be problematic for local farmers.

Early Environmental Consequences and Technological Solutions

Although the controversy over smelter smoke had largely subsided in the United Kingdom by the 1870s, the impact of expanding American smelter capacity generated new conflicts between agriculture interests and mining companies in this country. In the first decade of the twentieth

Figure 2.9. Open heap roasting in the Ducktown district, ca. 1900. Large heaps of sulfur-rich copper ores, partially covered, were ignited on a bed of wood and allowed to burn for several weeks while most of the combustible sulfur burned off. Reproduced with permission of the Edward Oscar Boak collection, Ducktown Basin Museum.

century, farmers and ranchers filed lawsuits against Montana, California, Utah, and Tennessee copper mines and smelters. The first well-documented court cases in the Tennessee district began in 1903 and 1904.[70] In both the consolidated *Swain v. the TCC-Cole v. DSCI* case and the *Madison v. DSCI* case, landowners lost their attempts to legally end open-heap roasting when Tennessee state courts ruled that the mining companies' benefit to the county outweighed their damages.[71] In the Madison ruling, the judge cited that twelve thousand jobs and a full half of the county's tax income were generated by the two companies.[72]

Concurrent with mounting and increasingly effective environmental law suits—including a case argued in front of the U.S. Supreme Court by the state of Georgia to end Tennessee open-heap roasting—both companies sought control over the roasting problem which, for them, was more of a production flow issue than an environmental nuisance. As early as 1899, DSCI experimented with pyritic smelting, the direct charging of green (unroasted) ores into its Herreshoff blast furnaces and using the heat of sulfur and iron

Figure 2.10. Isabella, Tennessee, ca. 1920. Effects of several decades of timber harvesting and open heap roasting have left this once heavily wooded southern landscape completely denuded of any vegetation. Reproduced with permission of the Ducktown Basin Museum.

oxidation to reduce the ore to a matte, thereby executing two smelting processes in one step. Because sulfur was flammable and sulfur and iron oxidation generated significant heat, roasting could be significantly sped up in a self-contained blast furnace by charging smaller amounts of ore and igniting the charge with little or no fuel. The combination of retained heat in the furnace and added heat from the oxidation and reduction processes in a fuelless *pyritic* or minimally fueled *partial pyritic* process meant that little fuel was ultimately needed for the roasting and first reduction stages, compared with open-heap roasting. Slag removed the iron oxide, leaving a 10%–20% first fusion copper matte, which would be cooled, broken up, and recharged with new ores and flux into a second furnace resulting in a 35%–50% matte that would be sold as is or sent to refining.[73]

By 1902, DSCI successfully arrived at a workable process and, in October 1903, ended open-heap roasting altogether and fully implemented partial pyritic smelting of all its ores.[74] By 1905, a short time after luring a smelting manager away from DSCI, TCC also began transitioning to partial pyritic smelting, abandoning all open-heap roasting, and constructing a 325-foot stack to disperse smoke over a wide area.[75] Like most American smelting innovations, pyritic smelting had its roots in Europe. Sticht reports in Peter's *Modern Copper Smelting* (1898) that Russian experiments in pyritic smelting date back to the 1860s, with significant advances by Englishman John Hollway at Rio Tinto, Spain in 1878.[76] Hofman reported successful applications in Montana, Colorado, and Tasmania by 1895[77].

In 1904, while both companies were studying or transitioning to partial pyritic smelting, the state of Georgia settled its initial lawsuit with the mining companies over the open-heap roasting process the mines were in the midst of ending anyway.[78] Despite the end of production bottlenecks and a symbolic end to open-heap roasting, the new methods did not alleviate the smoke problem—in fact, they ultimately lead to significant new lawsuits brought by the state of Georgia. According to Barclay, the smoke problem was exacerbated by the greater density of sulfur dioxide gas, significantly increased output, and new tall stacks that, like in Swansea, eliminated smoke locally but with higher air currents, disbursed smoke regionally, resulting in a more wide-ranging impact and a rise in distant lawsuits.[79] The copper district's proximity to Georgia, with geological features and prevailing winds directing smoke southeast and greater smoke dispersion from new stacks leading to greater cross-border impacts on Georgian farms and forests, created issues that copper company legal teams initially found relatively easy to maneuver around in Tennessee courts but would present greater challenges at the federal level.

A much more significant reaction to the smoke occurred in 1906, when the state of Georgia sued TCC, and later included DSCI, a second time in front of the U.S. Supreme Court, arguing that its sovereignty over its own land and air was further compromised by smoke from Tennessee copper production. The high court agreed and declared an injunction against smoke releases but delayed implementation pending a new technological solution to the problem that would not impact economic growth in either state.[80] In summarizing the majority opinion in the *State of Georgia v. TCC*, Justice Oliver Wendell Holmes wrote that Georgia wished "to enjoin the defendant Copper Companies from discharging noxious gas from their works in Tennessee over the plaintiff's territory. It alleges that in consequence of such a discharge a wholesale destruction of forests, orchards and crops is going on, and other injuries are done and threatened in five counties of the State."[81]

The Supreme Court ultimately postponed enforcement of the injunction pending the implementation of auxiliary sulfuric acid production plants at the smelter sites that would, in theory, eliminate sulfur and other "*poisons*" from the smoke. As early as 1872, August Raht had observed that Ducktown needed an acid recovery plant to remain economically viable, but sufficient technology and markets did not exist at that early date.[82] By the turn of the century, however, several proven European technologies existed and, in the midst of the 1906 Georgia lawsuit, DSCI attempted an experimental, but failed, contact acid plant. By 1909, however, both DSCI and TCC would successfully operate chamber acid plants.

Chamber sulfuric acid is produced by the oxidation of sulfur dioxide, obtained by burning sulfide ores in the presence of oxygen, its conversion into sulfur trioxide, and the combination of sulfur trioxide and water to form sulfuric acid (H_2SO_4). In 1908, TCC built a successful chamber acid plant at Copper Hill, which author Horace Stevens claimed was the largest ever built. The system routed blast furnace gases through dust chambers to a "nitrate apparatus," where nitrogen oxides were introduced into the hot gas. The nitrogen-enriched gas passed into the bottom of one of two octagonal, thirty-foot-in-diameter by fifty-foot-high Glover towers, where it encountered a combination of chamber acid and nitrous vitriol, an acid from the Gay-Lussac towers, trickling down from the top. The Glover tower encouraged nitrogen from the nitrous vitriol and steam from both acids to pass into the hot flue gas, thereby concentrating the chamber acid and denitrated nitrous vitriol. The combination, now essentially a simple concentrated sulfuric acid, was collected from the tower bottom and reused in the Gay-Lussac tower.[83]

The partially oxidized sulfur dioxide flue gas, nitrogen compounds, and some sulfur trioxide passed from the Glover tower through lead-lined flues to a 4.6 million square foot chamber for further oxidizing, hydrating, and condensing. Throughout the process, the quantity of sulfur dioxide gradually lessened from 7%, as the gases passed through several separate chambers converting to sulfur trioxide and acid, to less than 0.1% at the Gay-Lussac

Figure 2.11. Tennessee Copper Company chamber acid plant showing acid-forming chamber facility on the hilltop with acid storage tanks and rail cars in the middle ground, ca. 1920. Reproduced with permission of the Ducktown Basin Museum.

tower. Water mist was introduced into the acid chamber top and combined easily with the sulfur trioxide, forming sulfuric acid that condensed on lead sheets and was collected in pans below the chamber and sold commercially. The nitrogen-rich, desulfured gases passed from the chambers to one of ten Gay-Lussac towers, which trickled concentrated denitrated sulfuric acid from the Glover tower downward, absorbing up to 86% of the remaining nitrogen in the up-flowing desulfured gas. The now nitrogen-rich sulfuric acid, or nitrous vitriol, was collected from the bottom of the Gay-Lussac towers and used in the Glover tower to introduce nitrogen compounds to the flue gases. The desulfured and now mostly denitrated gases passed from the system and were vented into the atmosphere[84] (see figures 2.11–2.12).

Although the TCC acid plant was the largest in the world at the time, none of the prime processes it used had American roots. French chemist Joseph-Louis Gay-Lussac invented the Gay-Lussac tower in 1823

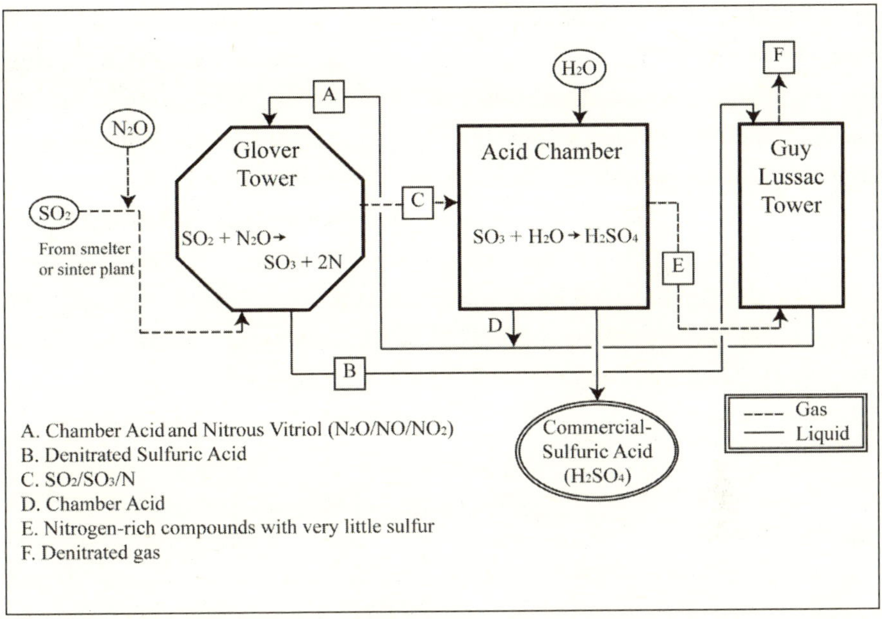

Figure 2.12. Chamber acid process for sulfuric acid production as used in Tennessee from the 1910s to 2000. Based on descriptions from Andrew Miller Fairlie, *Sulfuric Acid Manufacture,* 1936.

but found little use for his device until it was paired with the Glover tower, invented by Englishman John Glover in 1859.[85] Both sets of towers were lead-lined to keep the acid from deteriorating any structural elements, then lined with fire-brick and packed with flint, coke, broken wine bottles, or any other chemically neutral barrier to slow the ascent of smoke and the descent of acid.[86] Small lead chambers were first created in England in 1746 for sulfuric acid production with improvements in the late eighteenth century.[87] Contact acid production, a more complex alternative, developed in Britain in 1831 for converting SO_2 into SO_3 by drawing sulfur dioxide and air into a tube filled with heated platinum—or, later, vanadium oxide—that, like the nitrogen in the chamber process, encouraged the oxidation of sulfur dioxide.[88]

The key to successful sulfuric acid sales in the early twentieth century was the overproduction of agricultural lands, especially Southern cotton fields, and the need to renew them with fertilizers made with acid. As soon as DSCI's 1909 chamber acid plant went into operation, the company made more money from sulfur recovery and acid production than from copper production, relegating copper to a by-product of acid production.[89] As soon as TCC's plant began operations, the company signed a ten-year contract with International Agriculture Corporation for its entire output of sulfuric acid.[90] In 1909, 53% of all sulfuric acid was used in agriculture fertilizers, which dropped to 28% in 1918 when, during World War I, explosives took 36%, and the steel, petroleum refining, and chemical/drug industries used about 9% each. Through the mid-1930s, the fertilizer, petroleum, chemicals, coal, and steel industries were the top five users, respectively.[91]

Other Smelting Technologies

Completing the smelting process, TCC by 1911 and DSCI by 1920, adopted Bessemer-style converters for the refining of partial pyritic-smelted copper matte. Until 1920, DSCI sold all of its copper as high-grade matte, which required its customers to complete the refining process. TCC, with its new high-capacity Copperhill smelter works, likely installed its first converters in 1904.[92] By 1911, the company operated four converter stands with fifteen shells that produced a 99.4%-pure copper that TCC sold mainly in Europe as blister copper, although some copper pigs were sold to Baltimore Copper Smelting and Rolling Company, a later incarnation of the 1860 Baltimore Copper Company, for electrolytic refining.[93]

Although historian Barclay and engineers H. O. Hofman and Charles Christensen recognize that August Raht contributed to early copper refining by using forced air to oxidize iron and sulfur, most credit M. Manhès's work in Lyon, France, where he had successfully created Bessemer blister copper from matte in 1880.[94] Franklin Farrel built the first U.S. Bessemer copper plant at the Parrot Smelter in Butte, Montana in 1883. By 1885, the refining process involved charging a Bessemer converter with molten 40%–50% copper matte and blowing air through tuyeres into the copper bath, oxidizing the last iron that, in combination with siliceous materials in the converter lining, formed slag. Any remaining sulfur in the copper was also oxidized and vented.[95]

The only other significant smelting developments in the Tennessee copper basin occurred in the early 1930s, when both TCC and DSCI began phasing out their blast furnaces. By the 1920s, the copper content of the region's ores had dropped, contributing to a significant need for enhanced concentration before the materials could be smelted. In 1920, the same year it installed its first converter, DSCI built the first flotation plant in the district. Flotation-concentrated ore is not smeltable in a blast furnace, and the company sold this processed copper ore simply as high-grade concentrate. In 1925, DSCI reorganized as the Ducktown Chemical and Iron Company (DCI), and, after adding a second flotation plant in 1927, built an iron sinter plant. This new facility, which recovered significant quantities of iron from its ores, also permitted DCI for the first time in thirty-six years of calling itself an "iron" company, to actually produce and sell iron.[96] In 1931, DCI officially ended blast furnace operations, and all recovered copper was sold as concentrate, while the acid plant produced acid from the sulfur-rich iron sinter plant gases.[97] By 1936, the company's ore reserves were dwindling; it could not compete economically with its much larger neighbor and, amid the Depression, sold its sites and resources to TCC, leaving the latter firm standing as the sole producer in the area.

In the 1920s, TCC also began a program to eliminate its blast furnaces. Under the "all-milling" plan, the company built a flotation plant and a four-roaster iron sinter plant in 1923. By the mid-30s, TCC had completed its transition, which in 1936 included the incorporation of the already "all-milling" DCI facilities. With a second acid plant built at the start of WWI, new iron sinter plants constructed in the early 1930s, and a U.S. government–built acid plant constructed on its property during WWII, TCC had successfully diversified. It thus sold copper, different grades of acid, iron sinter, copper sulfates, fungicides, insecticides, zinc concentrates, organic chemicals, sodium hydroxide, ferric sulfate, and granulated slag.[98] Despite the apparent success of the "all-milling" program, TCC built a new reverberatory furnace at its Copper Hill smelter in 1940 and returned to smelting and refining copper.[99]

Montana's Richest Hill and Giant Smelters

The Michigan copper smelting industry experimented with advanced techniques, but in the end smelted its copper in basically the same way it had at the beginning, because Michigan ores neither warranted nor permitted great infusions of new technology. Tennessee's high-sulfur, iron-rich ores led to the early twentieth-century adoption of pyritic smelting and creation of colossal acid plants. Montana, on the other hand, not only advanced the scale of smelting, but also developed and refined existing European technologies—many of which made their way back to Michigan, Tennessee, and to the burgeoning new districts in Utah and Arizona.

As with Ducktown, gold prospectors first explored the Butte area. Unlike the Tennessee district, however, Montana prospectors discovered gold in placer deposits along streambeds and bottomlands. They established a mining camp in 1864, built the first regional smelter in 1866, and constructed a second furnace in 1868. By 1869, however, most of the gold placers were worked out.[100] Although a few quartz mines and mills remained open, Butte was practically abandoned until 1875 when miners discovered silver ore lodes several hundred feet below the surface. Silver proved to be a slightly more plentiful metal in the district than gold, and twelve mines operated in 1877.[101] Silver mining also led to other significant changes in the region. The burgeoning industry required a metallurgical infrastructure, and several mines built mills and smelters before the railroad reached the region in 1881. More importantly, one hundred feet below the silver lodes sat some

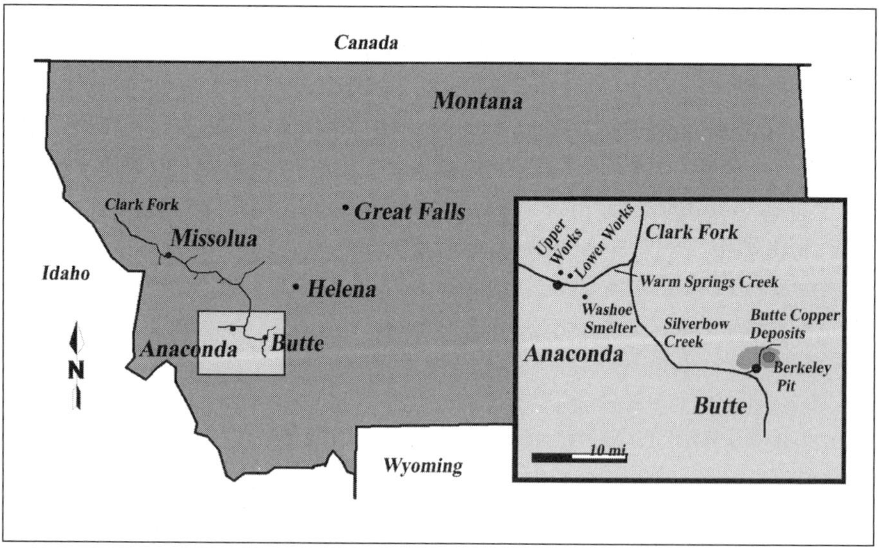

Figure 2.13. Butte-Anaconda copper district, Montana. By the author.

of the richest copper veins in the U.S.[102] Although copper was continually recovered in small quantities from the 1860s, the district remained primarily a precious metal mining camp through the early 1880s (see figure 2.13).

Early on all Montana copper ores were simply concentrated. The mines tried to import similar Cornish stamps as those used in Michigan, but the hard chalcocite (Cu_2S)—the region's primarily ore—was ultimately too brittle for stamping and led to significant milling losses.[103] Eventually the region settled on rollers, crushers, and water separation to process the ore and eliminate gangue, and initially sent high-content ores and concentrates to Swansea, Freiberg (Germany), and Baltimore for smelting and refining.[104] However, like all other copper districts, the rich ores soon gave way to poorer ores and mining companies introduced matte smelting and sent the partially smelted copper to Baltimore, Swansea, and Black Hawk, Colorado for finishing and refining. Most of the early 1880s copper smelters operated on Welsh principles with some combination of forty-five-foot "long-hearth" hand-rabbled reverberatory roasters and eighteen-foot reverberatory matting furnaces.[105]

William Clark's Colorado and Montana Smelter, in partnership with the Boston and Colorado Smelting Company from Black Hawk, Colorado, opened the first dedicated copper smelter in Butte in 1879. Following the opening of the Montana Copper Company's Colusa Smelter in 1880, the Parrot Smelter opened in 1881. Sometime after 1881, the founder of the original Colusa smelter started the Bell Smelter, featuring the first successful blast furnaces in the district for copper matte smelting.[106] In 1884, William Clark built a second smelter, this one called Clark's Colusa Smelter to distinguish it from the Montana Company's Colusa.[107]

In 1887, an East Coast capitalist formed the Boston and Montana Copper and Silver Mining Company and bought several properties from William Clark including "Clark's Colusa" mine and its smelter, which was renamed the "Lower Works" by the new owners. In 1888, the Boston and Montana purchased the Montana Copper Company's Colusa smelter, renamed it the "Upper Works," and ran both sites until it opened the Great Falls Reduction Works, 150 miles north of Butte, in 1893 and closed its two local smelters. A second Boston group formed the Boston and Butte Mining Company (1888) whose smelter, in 1890 successfully lengthened reverberatory matting furnaces to fifty feet. In 1887, William Clark bought the 1883 Butte Reduction Works, and operated the smelter until 1910 when it was purchased by the expanding Anaconda Copper Mining Company and closed just like the Colorado and Montana, Parrot, and 1892 Montana Ore Purchasing Co.[108]

In 1881, the second year that Irish immigrant Marcus Daly ran the Anaconda Mine in Butte, he returned a small profit from silver mining. In 1882, just as the Anaconda's silver lode ran out, a narrow, but promising, copper vein appeared several hundred feet below the surface.[109] By 1883, Daly was working an exceptionally rich copper sulfide vein, measuring between fifty and one hundred feet wide and assaying between 12% and 50% copper. In addition to encouraging significant improvements and expansions in the mine, Daly convinced his investors to build a captive smelter, not in Butte—where, he believed, land was too expensive and water too difficult to come by—but at a site on Warm Springs Creek, twenty-six miles west, that would eventually be named after the Anaconda Mine.[110]

The first Anaconda smelter, opened in 1884, included one of the world's largest concentrating mills, thirty-four hand-rabbled long-hearth reverberatory roasters, twenty-six matting furnaces, and two seventy-ton blast furnaces with a total site capacity of 450 tons of processed ore per day, with all matte initially shipped to Swansea for refining.[111] By 1886, the capacity of the smelter complex had increased to one thousand tons of ore per day, and Daly had replaced the labor-intensive hand-rabbled reverberatory furnaces with forty new Bruckner Roasters. In 1887, due to technological and space limitations of the works, Daly built a second smelter in Anaconda, which he designated the "Lower Works," and renamed the original smelter the "Upper Works."

The Lower Works burned down in 1889, and Daly rebuilt and expanded it—and increased the company's total capacity to three thousand tons of ore per day, adding ninety-six Bruckner roasters, twenty-eight matting furnaces, four Wethey roasters, eight McDougall roasters, six eight-foot blast furnaces, and three tall stacks to vent the works. In a move to further retain all capital from copper production and to complete refining in Anaconda, he installed a fifteen-stall upright Bessemer converter plant at the Upper Works, built the nation's third electrolytic refining plant between the two works in the mid 1890s, and added a converter plant to the Lower

Figure 2.14. Anaconda upper works smelter with lower works stacks in the background and electrolytic plant in the middle ground off to the right, ca. 1895. Reproduced with permission of the Butte-Silver Bow Archives.

Works.[112] The combined capacity of both works reached four thousand tons of ore per day[113] (see figure 2.14). With the new converters and refiners, copper shipped from Anaconda left at 99.97% pure, challenging the purity of Michigan Lake copper.

Although electrolysis was discussed above in conjunction with Calumet and Hecla's Buffalo Smelter, and Bessemer converting with the Tennessee Copper Company plant, both processes were refined in Montana before they were installed in either Michigan or Tennessee. The first U.S. electrolysis plant was built in Baltimore and the fourth in Buffalo; but the first plant directly associated with a copper mine was developed by Michigander Frank Klepetko for the Boston and Montana smelting works at Great Falls in 1893. Similar to the Buffalo site, Great Falls—with its new hydroelectric plant—offered great potential for inexpensive electricity and, with increasing debate over smelter smoke in Butte, the Boston and Montana Company built a large smelter and refinery at the falls and closed its Upper and Lower Butte works.[114]

The new 1893 Boston and Montana Great Falls smelter included twenty-four Bruckner roasters, eight reverberatory matte furnaces, and two five-ton upright converters. By 1902, the site included eighteen six-hearth McDougall-Evans-Klepetko roasters, five gas-fueled matting furnaces, five blast-furnaces, twelve upright converters, one anode-casting reverberatory furnace with a Walker Caster, a 312-tank electrolytic plant, and four reverberatory refining furnaces and was second in district production only to Anaconda, which by 1910 would purchase the Boston and Montana Company and the Great Falls refinery.[115]

According to Jonathan Leitner, the ability to home-refine Montana's copper matte that previously had been sent primarily to the United Kingdom proved to be the decisive blow to the already declining British copper industry.[116] Sharp declines in British production occurred in the 1890s as both Great Falls and Anaconda smelters added electrolysis, and again in the 1910s when Anaconda's capacity was greatly increased[117] (see figure 2.1).

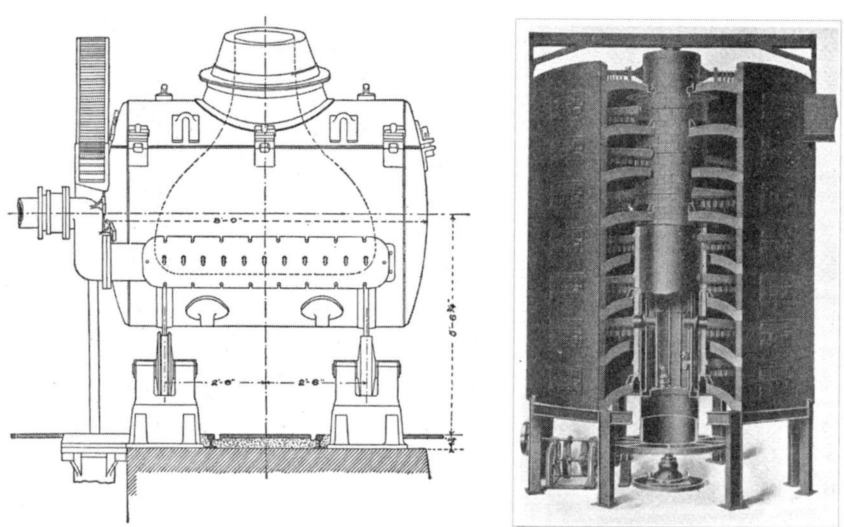

Figure 2.15. Upright bessemer converters for refining smelted copper matte from the reverberatory furnace (left) and a typical McDougall-style Wedge Roaster 33' high used to burn sulfur out of sulfide copper ores using the heat of sulfur oxidation to propel the process (right). Adapted from Peters, *Modern Copper Smelting,* 1898, and Hofman, *Metallurgy of Copper,* 1924.

TECHNOLOGICAL ADVANCES IN MONTANA

The Parrot smelter built the first United States Bessemer converter plant specifically designed for copper in 1884. It included three "egg-shaped" upright converters and a small cupola for remelting intermediate mattes. Despite two very different general arrangements, upright and horizontal, all converters operated essentially the same and, according to Laist, outside of modest improvements in charging and capacity and improved lining materials that decreased the frequency of lining changes from every six or seven charges to once every few years, few significant technological changes occurred with them through the 1950s[118] (see figure 2.15).

Following system enhancements in refining, Montana smelters turned to improving roasting. Like Tennessee pyritic smelting, roasting improvements were executed largely to improve production flow and reduce labor requirements rather than to curb or control smoke emissions. The first attempts at improved roasting were partially successful efforts to mechanically rabble ore in modified reverberatory furnaces. The first radical design change came with the Pearce-Turret roaster, whose thirty-six-diameter circular hearth had mechanical rabbles that rotated around a central shaft. The device had up to six circular stacked shelves to increase production, but its poor operating performance limited its wide-spread use.[119] The much simpler Bruckner Cylinder, a large horizontal cylindrical tub that rotated on rollers and tumbled ores over each other, saw heavy use, but its separate firebox used considerable amounts of fuel.

The most successful of the new roasters, however, adopted the upright cylindrical rotating circular shelf concept of the Pearce-Turret. The new devices were first developed by McDougall in Liverpool, England in 1870, and featured several stacked circular shelves, but most, however, were not larger than ten to fifteen feet in diameter. In the U.S., successful adaptations included the Wedge, Herreshoff, and the Evans-Klepetko developed at Great Falls. The McDougall-Evans-Klepetko married multiple-shelf circular hearths with a pyritic process in an enclosed and insulated circular tower. It used the heat of sulfur and iron oxidation to roast the ores on separate, nonrotating, enclosed shelves, with each shelf essentially becoming a separate hearth. Mechanical rabbles rotated by a central shaft directed ores to chutes that dropped them onto the next lower shelf. Without a separate firebox, the roasters took up considerably less floor space, required fewer workers to tend, and consumed virtually no fuel[120] (see figure 2.15).

Not all Montana smelters installed new-style roasters in the late nineteenth century. Not only were the hand-rabbled, hand-charged long-hearth reverberatory roasters still used, but also the smoke from so many smelting sites, some of which still heap-roasted large pieces or even stall-roasted ore concentrates, taxed the residents of the growing town of Butte. As early as 1885, citizens complained about the smoke destroying wild flowers and strawberries—but to no avail. Attempts to control the smoke or to dissipate it with short stacks largely failed, especially as output and heap roasting continued to increase through the 1880s as Montana surpassed Michigan's copper production. In 1891, Butte had had enough and passed an antismoke ordinance that ended all heap roasting and required all in-town smelters to erect a minimum seventy-five-foot-tall stack.[121]

The Dominance of Anaconda

By 1899, Marcus Daly's acquisition of other high-output Butte mines, and the limitations of the existing Upper and Lower smelting works in Anaconda, drove the need for smelter expansion.[122] Daly authorized $20 million for the project and hired Frank Klepetko from Great Falls to design and build a completely new facility that would become the Washoe smelter.[123] Daly directed Klepetko to design a single complex using the latest equipment to maximize efficiencies and to provide for future expansion in each department, but died before construction began in 1900. Continuing on, Klepetko selected a 230-acre site on an Anaconda hillside that ranged from 200 to 700 feet above the valley floor to take advantage of gravity—very similar to what he would design for the Michigan Smelter a couple of years later. The new, complete plant had concentrating, roasting, blast furnace and reverberatory smelting, converting, and electrolytic refining departments all linked by internal transportation with an original throughput of five thousand tons of ore daily. By then it was the largest copper smelting plant ever built, even eclipsing the earlier Anaconda smelters, and, by 1902, was consuming 750 tons of coal, 2,500 tons of lime flux, 778,000 kw of electricity, and 60 million gallons of water each day. By 1908, the expanded smelter processed 12,000 tons of ore and produced 600,000 pounds of copper and 9,000 tons of slag and tailings per day, all at a 39% cost reduction over the older works.[124] With the completion of the Washoe smelter, the Upper and Lower works became known simply as the "Old Works" and shortly thereafter were torn down. Over time, the abandoned site became an easily accessible dump.

According to Laist, few nineteenth century technological changes in smelting rivaled Anaconda's effort to radically increase the size of its furnaces. In 1904, Anaconda Copper Mining Company (ACM) took two of its 1901, fifty-foot-long reverberatory furnaces and joined them back to back with a single firebox at one end and single flue at the other, creating—in effect—the largest reverberatory furnace ever built. With such a large size, the Washoe smelters discovered that they could continuously charge the furnace at the firebox end and that the latent heat of the furnace "greatly accelerated fusion," effecting a 50% reduction in fuel consumed per ton of copper smelted with little increase in personnel needs. The new furnaces, now producing 250 tons of copper per day each, became the standard in the industry and were copied in Utah, Arizona, Nevada, and Mexico.[125]

Echoing the changes in reverberatory smelting at the turn of the century, blast furnaces, which did not undergo radical operational changes, simply got bigger as well. While the bosh distance between blast furnace tuyeres never got much wider than 57" to ensure adequate air pressure, the average furnace length grew initially from a 36" cylinder to an 8' rectangular bank. The biggest blast furnace at Anaconda appeared in 1906. It was 87' long and could process 2,500 tons of ore per day.[126]

By 1910, Anaconda's gigantic smelter complex operated sixty-four Evans-Klepetko roasters, eight reverberatory furnaces over 100' long, three Mathewson 87' blast furnaces, one briquetting plant, and eight 12' horizontal converters. Its concentrated ores were smelted in the large reverberatory furnaces for 45% copper matte. The matte was sent to the converter department and refined to 99% pure blister copper, remelted in a smaller reverberatory furnace, and cast into anodes that were taken to the electrolysis department and purified to 99.97%. The blast furnaces were charged with coarse ores, concentrate, reverberatory and converter slag, and ten-pound concentrate briquettes from the briquetting plant.[127] The entire smelter process generated between 3,000–4,500 tons of slag daily, which was discharged into flumes of flowing water to granulate it, then sluiced to dewatering ponds, which eventually grew into a single colossal waste pile that the company used from 1902–1976[128] (see figure 2.16–2.17).

Figure 2.16. Anaconda Washoe smelter, ca. 1950s, with separate concentrating, roasting, reverberatory furnace, zinc, leaching, and blast furnace buildings on the hillside and arsenic recovery, dust collection, and smoke stack on the hilltop linked by a flue. Reproduced with permission of the Marcus Daly Historical Society.

Environmental Problems and Responses

In addition to concentrator tailings and smelter slag, the site generated considerable smoke. Almost from the beginning of smelting at the Washoe, regional farmers complained about both the smoke problem and the damage to crops and livestock. In response to early complaints, Anaconda replaced four smaller stacks with a single three-hundred-foot stack on a plateau seven hundred feet above the valley floor in 1903. The stack was connected to the rest of the site by a 2,300' flue system designed to cool the smoke, catch dust, and recover much of the arsenic before it reached the top of the stack. The stack complex included a roasting furnace and small reverberatory furnace to process arsenical dust and a series of long narrow rooms, called kitchens, to encourage further arsenic condensation and recovery.[129]

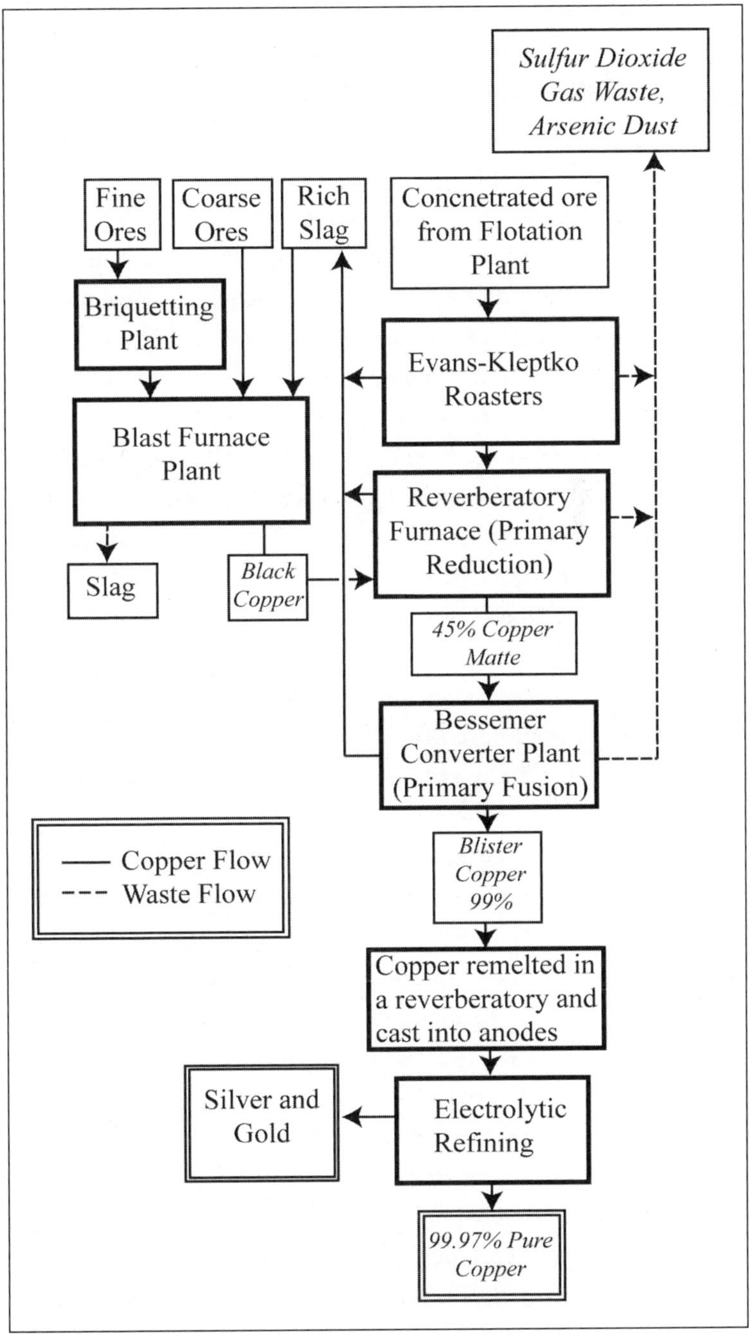

Figure 2.17. Montana smelting practices, ca. 1905. Based on descriptions in Brian Shovers and others, *Butte & Anaconda Revisited,* 1991.

Unimpressed and still suffering smoke problems despite the new stack, local farmers brought a lawsuit in 1905 against Anaconda's Washoe smelter, claiming continued harm was occuring to crops and livestock. Despite admitting to some damages and agreeing to pay modest penalties, Anaconda claimed that the farmers grossly overestimated their financial losses. In his ruling, federal district judge Oliver T. Crane agreed with the Tennessee state judge in the Madison case, maintaining that the benefits from smelting outweighed the damages to farmers, who relied on mining and smelter workers to purchase their goods. He awarded the farmers a meager $350 in damages.[130] In 1910, following investigations that had begun under President Roosevelt before he left office, the United States attorney general sued Anaconda for damages to federal lands with a suit modeled after the successful *State of Georgia v. TCC* case, and proposed a similar acid plant solution.[131] Unfortunately, an economic demand for sulfuric acid and fertilizer did not exist at the time in the West, and Anaconda agreed to independent monitoring by the Anaconda Smelter Smoke Commission, instituted by the lawsuit, and to equipment upgrades recommended by the commission, such as new Cottrell precipitators.[132] Despite poor potential for fertilizer sales regionally, the company added a one-hundred-ton chamber sulfuric acid plant in 1914 to recover sulfur from the smoke, possibly seeing potential sales for munitions at the start of World War I, and expanded its capacity to 215 tons per day in 1919.[133]

In 1914, Anaconda bought the International Smelting and Refining Company, expanding its operations into Arizona, Utah, Indiana, New Jersey, and Mexico. In 1915, Anaconda, which had been operating under the holding company Amalgamated Copper Company, returned with other Amalgamated assets and firms to a single primary corporation, the Anaconda Copper Mining Company (ACM), in an effort to economize management and fight lawsuits more effectively. In 1917, the smelting works expanded to process all the copper produced from new Montana mines and included one-hundred roasters, eleven blast furnaces, eight reverberatory furnaces, and seven converters. The increased capacity, however, overwhelmed the existing dust collection system by an estimated 50%. Responding to new recommendations of the Anaconda Smelter Smoke Commission and to the need for increased handling capacity, ACM constructed not just a larger smoke stack, but also, as locals boast, the tallest masonry structure ever built. The new stack measured 585' tall, sixty feet across the top, and seventy-five feet across the bottom (see figure 3.14). It featured a new Cottrell electrostatic plant that used an electrical current to more effectively precipitate and capture 90% of the arsenic from the flue gases.[134]

Twentieth-Century Evolution of Anaconda

Other new processes and improvements at the Washoe works included a 1915 plant that, through 1972, recovered zinc from Butte ores, and a 1915 flotation plant to concentrate its poorer ores like smelters in Tennessee and Michigan as the company started phasing out its blast furnaces. ACM also operated a brick plant from ca.1920 to 1980, a fertilizer plant that used acid from the chamber plant to create fertilizer from 1923 to 1961, and a ferro-manganese plant from 1941 until the 1950s. In 1933 ACM installed waste heat boilers on its reverberatory furnaces and, like the Michigan Smelter, used the steam to drive electrical generators that powered electric motors, ultimately producing up to 30% of the site's mechanical energy.[135]

Also in 1923, in a move to expand operations, diversify, and more fully vertically integrate as its Butte mine values and Montana ore reserves started to decline, Anaconda purchased the Chuquicamata mine in Chile, which worked one of the world's richest porphyry deposits, and soon after bought Chile's El Teniente and Potrerillos mines as well.[136] Anaconda would ultimately purchase brass rolling mills, copper foundries, uranium mines, aluminum, lead, and zinc works, and a host of other end-product manufactories in an attempt to reach complete vertical integration and become one of the largest nonferrous companies on the planet.[137]

As with all mine sites, the rich ores on which the Montana district was founded declined and continued to do so through the early twentieth century. Following a failed attempt at underground block caving in the early 1950s, Anaconda turned to open-pit mining in Butte and began operating the Berkeley Pit in 1955. This site was a deep new surface mine created to excavate the large low-grade copper deposit that sat below portions of the city of Butte and other district cities but was too lean to economically mine any other way. The pit not only forced the relocation of several neighborhoods and the destruction of popular parks, entertainment venues, and more than a dozen underground mines, but it also destroyed the late nineteenth-century smelting district that once housed the Boston and Butte smelter, Montana Ore Purchasing Company smelter, and Boston and Montana Upper and Lower works.[138] Although public outcry about the loss of town sites in the 1950s was muted by the promise of jobs and extending the life of the declining mining district, by the 1970s, talk of pit expansion and the threatened loss of Butte's central business district prompted the first elements of an organized protest movement. This locally organized movement would play a major role in redefining the city over the next thirty years, especially as the pit became the national archetype of mining-related environmental disasters. In 1976, the Atlantic Richfield Oil Company (ARCO) bought Anaconda after losses stemming from the 1973 expropriation of ACM's Chilean operations by the new government in Chile had significantly devalued the mining company. Lastly, although it replaced its reverberatory furnaces with electric furnaces to cut emissions in the 1970s, in 1980, fearing greater pollution liability, Anaconda closed its smelter for good.[139]

Important advances made in mining and milling techniques in the late nineteenth and early twentieth centuries, including open pit excavation and the extensive use of flotation, permitted the development of significant porphyry deposits in Utah and Arizona and new smelters, which were built at or near the mines. New copper lodes were also opened in Alaska, Nevada, and New Mexico and stand-alone smelters operated in Tacoma, Washington, and El Paso, Texas, to process much of these new ores. Despite accelerated production and consumption, few major technological changes occurred in copper smelting following the general demise of the blast furnace in the 1920s, until the wider use of electric furnaces in the 1970s and flash smelting in the 1980s.[140]

American copper smelting technology had evolved from relatively small-scale European-styled systems into some of the largest and most extensive metallurgical processes on the globe. Although European smelters developed technologies to smelt and refine copper over many centuries, their greatest technological advances came late—at a time of declining European mine production. The United States, however, with rapidly expanding electrical applications that ushered in the modern world, created unprecedented demand for copper. To meet the demand, miners and smelters adapted old and new European technologies to produce copper faster and cheaper—and to make available the extensive deposits of very low-grade ore that, prior to the end of the nineteenth century, the world had had little use for. Although none of the three sites discussed in this work mined very low grade porphyry ores, each adopted—and in some cases advanced or even perfected—technologies that allowed them first to profitably develop the richest ore deposits, then, with ever-advancing technologies, some of the leanest deposits in their respective regions.

These copper production advancements, which reflect important trends in American technological, social, labor and economic history, are at the very core of laws such as the National Historic Preservation Act (NHPA), designed to preserve the physical fabric of these momentous events and places. Each district has a rich collection of industrial, commercial, social, and domestic buildings as well as structures and landscapes that deserve and require proper conservation under the NHPA. Currently, the Michigan district houses a National Historical Park dedicated to copper mining, the Michigan and Montana regions support National Historic Landmark districts, and the Tennessee region supports several national register properties. All three have museums and heritage trails or other forms of public interpretation reflecting the importance of these places to their communities and to national history, but each district has also had to confront the environmental legacies of these rich histories.

Although many histories of the copper companies and copper-producing regions explore the technological, social, and economic developments of their productive years, few, until the 1990s, fully considered the aftereffects of this production and the waste streams generated from ores with as little as 2% or less copper. Chapters 3, 4, and 5 of this book will examine the post-mining eras in these copper-producing regions when copper companies ceased, or very substantially reduced their mining and smelting operations. One prevalent legacy was the degraded environment, a product of a century or more of producing mine tailings, slag, acid rain, and deforestation. In some copper mining and smelting districts, the degradation was severe enough to get portions listed as "Superfund" sites, which involved federal environmental remediation efforts and oversight. Another legacy was, ironically, a growing interest in preserving elements of the same industrial heritage that had caused the environmental degradation and using it to promote heritage tourism to replace the now-lost industrial payrolls.

Chapter 3

Ducktown

> Observations were made on the condition of the vegetation in the region surrounding Ducktown and Copperhill, Tennessee in the neighborhood of the copper smelters during the month of September, 1913. . . . The area affected by the acute form of injury is probably larger than it was in 1905 and 1906. . . . In the worst affected area little herbage is found, and the hills look like a desert with a greatly eroded appearance. . . . Repeated acute injury of this sort kills the twigs of trees and finally brings about the death of whole trees.[1]
>
> —George Grant Hedgcock, 1914

> [T]he landscape opens up, and suddenly there is a vast, raw plain, cooking in the summer sun. No trees, no plants, no signs of life, just a great expanse of dry red-clay hills cut deep with massive gullies . . . so strange, so sudden, so out of place.[2]
>
> —Wilton Barnhardt, 1987

> "It's sort of sad. I've grown up with the painted cliffs and red hills and, in a curious way, they're quite lovely and unique. . . . It's our history and we're proud of it. . . . They ought to leave it to show what man can do to nature [and] what man can undo."[3]
>
> —Loraine Lee, 1987

April 11, 2008

The final part of the drive into Ducktown, Tennessee from Chattanooga is rather unremarkable, especially given the striking, thirty-mile climb that follows the Ocoee river gorge up from Cleveland (Tenn.). One would hardly know that this stretch of river was once classified as ecologically dead and not expected to fully recover for decades. Dead, however, is a relative term. The fall of the river over thirty miles powers three Tennessee Valley Authority hydroelectric plants, offers recreational kayakers and rafters occasional white water, and was the site of the 1996 Olympic paddling course. Fish and aquatic plants and insects, however, are sparse here. Heavy metals, arsenic, and acids washed into this watershed from over a century of copper mining and smelting operations in the Ducktown district. Some contamination resulted from the direct effluent of processing operations, but the majority came from serious and significant erosion caused by several decades of open sulfide ore roasting that, combined with mining-driven lumbering operations, left a completely denuded landscape of dirt hills and sediment-filled valleys.

Today, the Ducktown area is a bucolic, forested hilltop looking like any other small town in eastern Tennessee accessible by a four-lane highway exit ramp. Unlike Copperhill, its sister-city at the southern edge of the Copper Basin five miles south, Ducktown has a small commercial strip, a fast-food restaurant, and a hotel—but no railroad station for excursion train stops. Driving east on U.S. 74, the only hint of any sort of past industrial activity are signs bolted to chained gates that read "McPherson Mine" or "London Mill" or, one half-mile off the highway in downtown Ducktown, "Burra Burra Mine" and "Ducktown Basin Museum," the latter with its collection of historic buildings and equipment largely invisible from the road.

The imagery changes along state highway 68 as one heads south to Copperhill. The large flag-topped head frame of the Central Mine towers above the trees shortly before the "Smelters" gristmill store and fireworks stand. Farther south along the highway into Copperhill, however, the extensive, chaotic, partial remains of the Tennessee Copper Company smelter and acid plant vividly announce recent and not-so-recent industrial activity, while a quarter mile beyond the plant, quaint downtown Copperhill's copper-accented street lights, heritage signs, and mineral gift shops quietly suggest a past worth remembering.

A mere thirty-five years ago, however, the landscape of the Tennessee Copper Basin looked entirely different. Despite the cessation of open-air ore roasting by 1905, the area remained sharply denuded and moon-like, with large areas of very little vegetation as late 1973. Locals claimed that the fifty-square-mile basin was the only man-made landscape visible from space and among the myriad national articles and photographs about the district is a two-page *National Geographic* aerial image showing the sparsely populated built environment completely engulfed by totally barren red-sand hills.[4] Although an abomination to some, this landscape was mourned by many locals who grew up with it once it was gone. Like the once-belching smoke stacks of Pittsburgh's blast furnaces, the denuded landscape was a symbol of the region's prosperity and a significant component of its identity. Although attempts to preserve a small thirty-acre plot of barren hillside behind the museum have succumbed to successful revegetation and remediation efforts, preserving and celebrating the Basin's remaining mining heritage remains a key development strategy for Ducktown and Copperhill (see figures 3.1–3.2).

Shutdown

Through midcentury, the Tennessee Copper Company (TCC) added further plants in Copperhill to produce a variety of chemicals. In fact, its by-product operations were so successful that by 1950 copper was relegated to the third leading product manufactured after sulfuric acid and iron sinter.[5] In the 1950s, TCC's chemical line grew even further after it added an organic sulfonating plant in 1952 and a sodium hydrosulfite plant in 1956. In 1964, the company closed its last chamber acid plant and constructed its fifth contact acid facility at Copperhill to complement existing contact plants built in Isabella and Copperhill between 1929 and 1954.[6] The Ducktown Sulpher Copper and & Iron Company (DSC&I) dismantled and either scrapped or moved its Isabella chamber plants before being purchased by TCC in the 1930s, and TCC ultimately removed the remaining Isabella chamber and defunct contact plants in the 1970s.[7]

In 1963, in a move that foreshadowed later oil company takeovers of mineral companies, Cities Service Corporation, the Oklahoma-based predecessor of CITGO Petroleum, bought the Tennessee Copper Company, continued underground mining and chemical production, and in 1976 began open-pit mining northwest of Copperhill.[8] In 1982, the recently organized Tennessee Chemical Corporation (TChC)[9] acquired the Copperhill and Isabella processing plants and remaining active mine assets from Cities Service, while Occidental Petroleum purchased the rest of Cities Service,

Figure 3.1. Effects of remediation. The aerial photo of Ducktown on the left was taken in 1944; the photo on the right was taken in 2008 from the point marked on the photo on the left demonstrating significant changes in the landscape. 1944 photo reproduced with permission of the Ducktown Basin Museum. 2008 image by the author.

including its substantial oil refining and distribution network and defunct Tennessee copper mining and smelting sites. Occidental, however, did not hold onto the Cities Service assets very long, selling them to the Southland Corporation the following year. In 1987, Tennessee Chemical ended copper mining in the state and in 1989 declared bankruptcy.[10] In 1990, Boliden Intertrade, a Swiss corporation, purchased the Copperhill plants from bankruptcy and entered into an agreement with the U.S. EPA not to sue (or be sued for environmental damages of TCC or TChC) and continued operating the acid plant through 2000 and the organic chemical plant through 2008.[11] Properties not associated with Boliden's Copperhill operations remained in the bankruptcy estate of the Tennessee Chemical Corporation.[12]

When it stopped producing acid and significantly reduced its workforce, Boliden eliminated all but 300 of what, in the 1970s, had been as many as 2,000 jobs, resulting in a considerable drop in the local economic benefits from mining and chemical production. Where the 1906 *Madison vs. DSCI* lawsuit referred to 12,000 jobs associated directly or indirectly with mining and claimed that half the county's taxes were paid by DSCI and TCC, the U.S. government was the largest single employer in the county in 2001 and

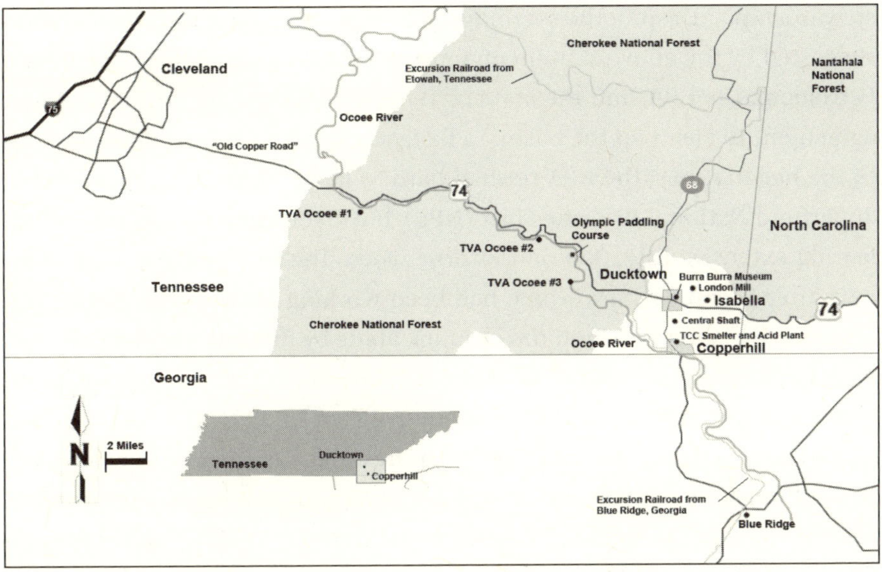

Figure 3.2. Map of Tennessee Copper Basin showing selected heritage and remediation sites. By the author.

controlled almost one-third of all land through the U.S. Forest Service. With inexpensive housing, open forest lands, and the Ocoee River area offering hiking, biking, rafting, and kayaking, tourism became the primary income generator for the county, a fact not lost on community leaders looking to create opportunities and accessible landscapes for short-term visitors.[13] Mining and agriculture, historically the two largest economic sectors in the county, had significantly lost influence by 2003, as locals focused efforts on developing a second home/retirement market to capitalize on the "environmental resources of the area."[14]

Although Occidental Petroleum sold Cities Service in 1983 to the Southland Company, which by 1990 had sold it to the national oil company of Venezuela, Occidental remained the last viable district stakeholder with historic liability.[15] Indeed, according to the 2009 website of Glenn Springs Holdings, an Occidental Petroleum environmental subsidiary company: "In 1987 the EPA began investigating the site [the Copper Basin] and documented its negative environmental impact. Since the only remaining stakeholder was Occidental Petroleum Corporation, Occidental agreed to remediate the site in good faith in keeping with the company's commitment to environmental stewardship."[16] Despite the seemingly effortless transition into remediation suggested by the above statement, fourteen years passed before Occidental Petroleum, the EPA, and the state of Tennessee executed a formal consent agreement to clean up the basin. In fact, according to federal officials close to the negotiations, the EPA pushed hard to have the region placed on the Superfund National Priorities List (NPL), but Occidental fought the listing, fearing extensive litigation and attorney fees. The two reached agreement only after Glenn Springs, which had been working in the basin since 1997, "voluntarily" agreed to fund the cleanup, abide by EPA rules and guidelines, and reimburse the EPA and the state of Tennessee for costs incurred. In an historic memorandum of understanding signed in 2001, the EPA, in turn, agreed to let Occidental and Glenn Springs manage the cleanup privately and to limit federal involvement to monitoring and oversight. The state's Voluntary Cleanup, Oversight and Assistance Program (VOAP) also kept Glenn Springs from potential long-term state liability as long as the state was satisfied with the cleanup.[17]

Early Reaction to Environmental Degradation and Response

It is true that citizens of Lower Wales and Southwest Montana complained about and sued copper smelters over lost vegetation and sick animals, and that smelter smoke court cases, newspaper accounts, and mining textbooks often referred to barren and contaminated landscapes surrounding smelters. Nonetheless, few regions experienced debilitating effects of landscape change as great as those of the Ducktown district. As noted in Chapter 2, the loss of woodlands surrounding Tennessee copper mining and smelting locations completely shut down the industry in 1878 and, following a resumption of mining and smelting in the 1890s and legal and technical efforts to reduce sulfur emissions, district companies and federal agencies began coordinating efforts to solve the landscape problem in the 1920s. Although Michigan copper regions harvested significant amounts of timber for mining operations and Montana smelter smoke led to considerable vegetation loss, efforts to combat these problems if they occured at all usually resulted in court orders or municipal actions ending the injurious practice or in presenting damage awards to affected parties—not to complete shutdowns. Tennessee observers chronicled the problem more completely, and companies took actions to remediate the problems well before any other district.

That this attention was given to remediation so early does not necessarily mean that Tennessee corporations were more environmentally responsible than other mining companies; rather, the Tennessee copper district operations caused significant problems outside the basin, especially noticeable after Franklin Roosevelt created the Tennessee Valley Authority (TVA) in 1933. The TVA's jurisdiction extended over the entire Tennessee River drainage, including Davis Mill Creek and North Potato Creek, both of which ran through the Copper District, fed Ocoee River hydroelectric plants, and ultimately emptied into the Tennessee River. Although the extraregional awareness created by TVA resulted in funding for studies and remediation efforts, the conservation movement actually began in the late1920s, and it is difficult to define what early drivers were present in Tennessee that didn't exist in other regions. Karen Daniels argues that the Supreme Court case that Georgia filed created regional awareness as early as 1907 and ignited the conservation movement, but court cases occurred in other areas, including Montana, Utah, and California that did not foster a sense of community responsibility or a need to remediate the aftereffects of smelting.[18]

Tennessee's landscape itself may have contributed to a stronger sense or perception of loss. The barren red-soil lunar landscape of the Ducktown district sat in stark contrast to the thickly wooded green Appalachian forests that visitors traversed to reach the district, strongly emphasizing the devastation. Similarly, southern residents in general identified more closely with their culture's agricultural roots than with the industrialization often associated with n*orthern* culture. For a short time in the 1930s, a group of Tennessee scholars and writers called "The "Fugitives" or "Southern Agrarians" sought to end creeping industrial materialism and return the South to traditional agrarian values. Although the group did not foster much change—and was significantly marginalized by the Depression—these scholars and writets published a controversial manifesto, *I'll Take My Stand: The South and the Agrarian Tradition (1930),* and reflected some of its generation's deeply seated beliefs.[19]

Thus landscape devastation in southeast Tennessee would have caused greater concern culturally than in areas with a more industrial culture. Further, Michigan mines and smelters, followed by Montana mines and smelters, each led global copper production, creating a stronger regional/statewide industrial identity—and even pride for the productive output of their respective districts. Big mines led to big production, greater population densities, and population diversity reflected in vibrant downtowns and mining colleges that continued as technological universities after the end of most mining. Tennessee mines and smelters never produced at the rate of those in either Michigan or Montana and had more visible environmental problems and landscape degradation. With strong agricultural identities, more state and regional residents were therefore likely to react to negative landscape change than industrial productivity. In any case, writers there (mostly nonlocals) described the landscape changes earlier than in other copper smelting districts, and companies tried to remediate the environmental effects earlier than those in other districts. Later local residents would try to preserve the desolate landscape as an element of their historic identity, earlier and more vocally than their counterparts in other copper regions.

A short, unscientific review of popular web-based twentieth-century imagery from the three districts also demonstrates key identity issues for the areas. Where regional Montana photographs and postcards show mining activities, smelting equipment, downtown buildings, and people, and regional Michigan photographs show miners, industrial structures, and shipping, popular Tennessee imagery is dominated by the denuded landscape. Both Montana and Michigan regions are depicted by images of the built environment. Tennessee is depicted mostly by pictures of environmental destruction.[20] Not until the Superfund era and greater environmental consciousness did popular views begin to focus on environmental decline in Montana, mostly at the Berkeley Pit, while in Michigan this awareness has still not occurred despite the Superfund cleanup.

The denuded Ducktown landscape was a direct result of mining and smelting operations, but M. –L. Quinn suggested twelve reasons why inherent landscape conditions exacerbated the devastation: From isolated mountains with poor transportation routes, to high sulfide and pyrite concentrations, with high rainfall, humidity, and temperature inversions, the environment itself was not well equipped to cope with significant mining and smelting operations.[21] Quinn continues: "Environmental susceptibility is an identified condition whereby the environment of a particular location, because of the location's natural features, is especially vulnerable (is predisposed) to injury from specific human activity. Susceptibility, therefore, is a function of a site's natural features, but the basis for its determination is the manner in which those features interact, or could interact, with a particular human activity."[22]

In any case, the estimated loss of fifty square miles of forest and vegetation was formally recorded in 1929, and in 1944 the TVA estimated that basin lands eroded at a rate of thirty-nine ac-ft/mi^2/year, while the rest of the Ocoee River drainage averaged only 0.44 ac-ft/mi^2/year. Further, by 1951, TVA had documented that in two separate five- and fifteen-acre parcels, erosion had consumed a total of 1.8 and 1.9 feet of topsoil and subsoil over seventeen years. All of that sediment, much of it contaminated with acids and heavy metals, found its way into the Ocoee River.[23]

Despite an early and clear understanding of both vegetation loss and the effects of soil erosion, especially during heavy rains, efforts to reforest the basin and control sediment loss did not begin in earnest until 1929, when the Ducktown Sulpher, Copper, and Iron Company (DSCI) sought advice from the U.S. Forest Service. With the support of the Department of Agriculture, DSCI planted and/or distributed three thousand trees in 1930 with the assistance of the University of Tennessee extension service, and continued this practice in subsequent years. Although no clear reason for the company-sponsored revegetation emerges from the literature, Edward Johnson cited a 1929 Forest Service report that included a discussion of financial returns from timber production based on market demands and wood quality.[24] Although Johnson does not state whether financial returns were the prime reason for tree planting or, in fact, a secondary benefit, the late-1920s economic decline in copper and acid production, which ultimately led to the 1936 acquisition of DSCI by TCC, may have forced the company to look for alternate means of income. In 1973, a Cities Service report reflecting historic corporate positions stated: "The land reclamation program carried out over more than 35 years by TCC and subsequently Cities Service has been carried out in the interest of better utilization of our natural resources (land) by putting it into productive forest land use as well as reducing possible siltation of surface waters in the basin. This program has been carried out as a Company Policy and has been in no way connected with legal or administratively imposed requirements from any governmental body or Agency."[25] Despite limited culpability or admission of responsibility, the seemingly defensive company posture may also have been influenced by political pressure, reactions to negative accounts of the landscape, problems with the TVA, or the threat of potential lawsuits.

Through the 1930s and 1940s, several groups, including the Tennessee Extension Service, U.S. Forest Service, U.S. Bureau of Plant Industry, Civilian Conservation Corps (CCC), Federal Soil Conservation Service, TVA, and universities in Tennessee, Georgia, and North Carolina, studied and worked to continue area reforesting efforts but with only partial success, because many early plants did not survive long enough to reproduce and establish new ground cover and forests.[26] Between 1939 and 1950, the TVA, which had done little direct work on the Copper Basin erosion problem in the 1930s despite significant siltation effecting its two (at the time) downstream hydroelectric plants, took the lead in research and, with an established CCC camp, in 1941 planted trees on the perimeter of the worst affected areas and experimented to determine which tree and shrub species grew best in the basin's acidic soils.[27] In 1943, TVA built a third hydroelectric plant on the Ocoee; this one, just a few miles downstream from the basin, included a large reservoir to stem the flow of sediments and heavy metals to its downstream plants.[28] According to Muncy, however, by 1973 sediment completely filled the Number Three hydro-plant reservoir, requiring occasional sluicing of sediments contaminated with heavy metals and acids downstream.[29]

TCC largely took control of the revegetation efforts from the TVA in the 1950s and hired a professional agriculturalist to continue its tree-planting program.[30] Through the 1970s, a variety of programs continued, each with some degree of success, as plot by plot, the 32,000 acres of denuded landscape shrank to 20,000 in 1943, 12,000 in 1986, and close to 1,500 in 2000.[31] Although successful experiments with the "miracle legume" kudzu in 1956[32] proved that the plant was a "competitive threat" to other flora,[33] most new plantings usually included some combination of loblolly pine, Japanese fleece flower, or weeping lovegrass. Each of these plants demonstrated resistance to district toxins, especially when augmented with new fertilizers, planting techniques, and, in the 1970s, rapid aerial seeding and feeding. Between 1950 and 1969, 4.5 million trees were planted, and between 1970 and 1986, 6.85 million.[34] By 1999, over seventeen million trees had been planted since the TVA-funded reforestation efforts had begun in

1941.[35] By 2008, most sources cited only one fully unvegetated thirty-acre parcel, intentionally left denuded behind the museum that, because of dangerous subsidence from a mine collapse, could not be entered to remove the volunteer trees successfully taking root.[36] Although nearly seventy-five years of plantings have significantly reduced the desert appearance, a 2008 drive through the basin reveals a few remaining bare spots.[37]

About the same time as heavy revegetation efforts began, community awareness and activism crystallized at a rally of the Tennessee Conservation League held in the Copper Basin on May 27, 1950, organized for the kickoff of a statewide "Keep Tennessee Green" campaign and to recognize the successful contributions of local industry. The speakers included Tennessee's governor, TCC's general manager, and regional historian R.E. Barclay, among others. Several participants focused on the success of early reforestation efforts in the district, which was then about 50% recovered.[38] Governor Browning remarked that, as a nation, "conservation of our natural resources must be our aim if we are to survive" and "that the human race is just 8 inches from starvation and one-half of this has gotten away from us," referring to lost topsoil in the basin and reflecting the dust bowl that had occurred just fifteen years earlier. TCC's general manager proclaimed that the copper company had only just begun remediating and pledged to "Keep Tennessee Green," while R. E. Barclay asserted, after extolling the contributions of the copper company to district employment and taxes, that communities could rebuild lost houses and cars, but that without plants, they would surely perish. Probably the most telling observation—and an early indication of changing directions in economic thought—came from C. P. Swan, state commissioner of conservation, who described how statewide tourism had become a $225 million dollar industry by 1949, third only to manufacturing and agriculture, and with better conserved and managed green spaces, Swan saw no reason why that that amount could not be doubled[39] (see figure 3.3).

Efforts to sway negative outside public opinion and to convince critics and potential tourists that the ongoing remediation was successful included a one thousand–acre bluegrass and clover-covered demonstration farm that supported a one hundred–head herd of "purebred Aberdeen-angus" cattle and a nine-hole golf course right in the heart of the district. These 1950s projects were designed mainly to serve as "proof that the soil

Figure 3.3. 1950 conservation rally in Copperhill, Tennessee, to kick off the "Keep Tennessee Green" campaign. Courtesy of the Tennessee State Library and Archives.

[wasn't] as dead as some of it looks" despite the fact that TCC lost money on their operations.[40]

Origins of Ducktown Heritage

In popular articles with such titles as "Ducktown-A Post War Challenge" (1943), "Healing the Scars" (1956), "Tennessee Valley Today" (1973), "Copper Basin Cover-up" (1983), "Death of Ducktown" (1987), "Industrial Destruction Reversed at Copper Basin" (2002), and "Up from the Mines in Tennessee" (2007), journalists for over sixty-five years made a point of describing Ducktown as a landscape mired in environmental conflict and asserting that only near superhuman efforts had ultimately resulted in a resounding remediation victory.[41] Although these authors represented mostly an outsider perspective about the remarkable turnaround in just fifty years, others questioned whether such a full restoration was of any value to district resident insiders.

M. –L. Quinn's 1992 article, "Should all Degraded Landscapes be Restored?" in *Land Degradation and Rehabilitation Journal* complemented four additional articles she had written on the historical and environmental forces that shaped the Tennessee copper mining district's landscape. In her *Land Degradation* article, she reflects on the desires of some locals who felt a sense of loss because the heritage landscape they grew up with changed so dramatically as the restoration process neared completion. Quinn argues that all reforestation efforts should stop (as of 1992) until the extant mining/smelting landscape could be surveyed and documented for its unique and historically important characteristics and to determine if remediation efforts conflicted with or diminished the cultural and historical value of the basin—similar to processes required by federal projects under Sections 106 and 110 of NHPA. Quinn further calls for independent historical assessments because, she argues, TVA and the U.S. Forest Service had been too zealous in their reforestation efforts and the EPA's mission of protection and remediation rendered it incapable of objective evaluation.[42]

Similar themes surfaced in Barnhart's and Dixon's Ducktown articles, but Quinn's notions, however, reflected a growing body of literature questioning the extent and value of some remediation efforts. Archeologist Don Hardesty wrote "Issues in Preserving Toxic Wastes" (2001) in the *Public Historian,* specifically questioning the value of remediating all toxic sites, especially those that have recoverable archeological data or heritage value and are not, of course, posing immediate threats to human life.[43] Similarly, David Robertson wrote in "Heaps of History" (2000) in the *Journal of Illinois History,* that despite the negative association with waste streams to outsiders, there may be greater community value than harm in not fully remediating those waste streams. Robertson describes the case of Toluca, Illinois, whose residents unanimously approved historic landmark status for "Jumbo," a massive coal mine waste pile.[44] Even more critically, Fredric Quivik questioned the authority of the EPA to ignore its National Historic Preservation Act obligations in sensitive cultural areas in "Integrating the Preservation of Cultural Resources with Remediation of Hazardous Materials: An Assessment of Superfund's Record" (2001) published in the *Public Historian.*[45]

The issue of preserving a portion of denuded basin amid efforts to fix the erosion problem drew considerable outside attention, especially as the revegetation efforts were proving so successful. Many district residents wanted to keep at least a part of their "beloved scar,"[46] and a community survey conducted in 2001 cited several residents who wished to preserve the "moon-scape."[47] Even as late as 2008, when most of the open but inaccessible thirty-acre plot behind the museum, initially kept barren as a reminder of earlier landscapes, was well on its way to volunteer reforestation, Ducktown's mayor still wanted, but didn't know how, to keep part of the small parcel free from trees to commemorate the region's history[48] (see figure 3.4).

The particular challenge for locals was to balance its heritage needs (and desires) while overcoming the real and perceived ill effects of the region's mining and smelting past and yet somehow tap into the "4.5 million stressed out Atlantans, just two hours south" as potential short-term, income-infusing visitors.[49] Unfortunately for the city of Ducktown, the 2008 north-bound

Figure 3.4. *Left*: The 30 acre parcel behind the Ducktown Basin Museum in 1950 looking northeast. *Right:* The same parcel in 2008 looking southeast. The plot, which was curiously treed in the 1950s, was intentionally devegetated to leave a reminder of the basin's once denuded landscape; however, because of the success of new species introduced, the parcel, which is now off limits because of the visible mine collapse, is reforesting itself naturally. *Left:* courtesy of the Tennessee State Library and Archives. *Right:* by the author.

excursion train from Blue Ridge, Georgia and the south-bound train from Etowah, Tennessee ran on existing tracks and stopped in Copperhill, completely bypassing Ducktown. So although Ducktown has the commercial strip and museum, it doesn't have the built-in tourism vehicle that Copperhill has, but other opportunities may exist with driving visitors who may be convinced to stay longer than the mandated stop during rail excursions. But that reality is yet to be seen. Ducktown thus struggles for regional identity and greater authenticity in its heritage planning effort to create a stronger tourism market.[50]

Like other rural mining districts, especially those that functioned for over a century and were in some state of economic decline, third- and forth-generation basin residents sought to commemorate their heritage and celebrate the people and places that gave their region distinction. In 1978, an impromptu mining exhibit began in a Ducktown grocery store with some bartered artifacts. It ultimately grew into a more formalized museum space where former workers and local residents gathered personal collections, stories, and photographs. Shortly after the new museum opened, Cities Service abandoned the Burra Burra Mine site in Ducktown, which had served as the TCC surface headquarters from 1899–1976. In 1982, Cities Service donated the mine site to the new Ducktown Basin Museum as an historical monument and exhibit space. The museum soon occupied ten buildings on seventeen acres, including the former office building, machine shop, and ore bin ruins and, in 1983, the site was nominated to the National Register of Historic Places.[51] In 1988, the museum transferred ownership of the site and artifacts to the state of Tennessee but continued to manage and interpret the mine site, artifacts, and local history through a separate nonprofit organization.[52] Although it interprets environmental conditions and copper milling and smelting, the museum is primarily a mining museum and, with the exception of the attempt to preserve the small smelter-denuded lot behind the museum, does not fully address the complex and significant (both positive and negative) smelting operations of the district. Very similar mining museums operate in Butte, Montana and Hancock, Michigan, but both of these districts have made other, more formalized efforts to include part of the built smelting landscape.

The only other significant heritage activities to take place before the remediation efforts were multiple national register nominations for districts and sites in the basin. In 1992, the Southeast Tennessee Development District nominated the Ducktown and Copperhill downtown business districts, the last-standing and regionally unique Central Mine head frame, and two individual nonindustrial buildings to the National Register.[53] With the exception of the inaccessible but highly visible Central head frame, all nominated structures and districts, including the 1983-nominated museum site, were publicly accessible. The more significant industrial site, the Copperhill smelter and acid and organic plants, however, was left off. Local authorities suggest that planners considered listing the site, but were discouraged by the smelter site's Swiss operator, possibly because of unfounded fears of federal restrictions on site use or simply of the focused attention of a nomination. Further the Copper Hill plant, with all of its disarray and destructive appearance, never quite fit well with the clean, controlled, and "attractive" appearance of its National Register–listed neighbors and shopping districts. In fact, as heritage planning and remediation efforts continued after the memorandum of understanding was signed in 2001, the still-active Copperhill site continued to be excluded from heritage planning.[54]

Superfund

Although the most visible effects of smelting operations in the Ducktown district were remediated by extensive tree planting over a seventy-five-year period, few articles or studies mentioned the extent of "invisible" toxins remaining in the district until the advent of the Clean Water Act (1970), Resource Conservation and Recovery Act (1976), and Superfund (1980). From 1970 to 2001, several studies examined the remains of mining and smelting works in the district and attempted to quantify the extent of environmental damage. At many locations, heavy metals including copper, lead, iron, manganese, aluminum, and zinc, as well as acids, PCBs, and other chemicals, remained in the soils from a century of operations. For example, one report listed the following contaminants, many carcinogenic, in one soil sample taken near the Isabella plant in 2004: 2-Methylnaphthalene, Benzo(a)anthracene, Benzo(a)pyrene, Benzo(b)fluoranthene, Benzo(g,h,i)perylene, Benzo(k)fluoranthene, Chrysene, Fluoranthene, Indeno(1,2,3-cd)pyrene, Naphthalene, Phenanthrene, Pyrene, Cobalt, Copper, Iron, Lead, Manganese, Selenium, and Zinc.[55]

Under CERCLA, the EPA retains the authority "to pursue long-term remedial response actions that permanently and significantly reduce the dangers associated with releases or threats of releases of hazardous substances."[56] However, before it can act and spend federal money, the EPA needs first to verify the extent of the damage, then initiate response actions and pursue those ultimately responsible for paying for it. The following passage comes from the EPA's description of the Copper Basin on its website (accessed March 13, 2009):

> Mining and related activities have resulted in the environmental degradation of portions of the Copper Basin, including the North Potato Creek Watershed, the Davis Mill Creek Watershed, and parts of the Ocoee River. Waste materials from mining and processing activities remain as sources of contaminants in the form of acidic drainage and high levels of metals in the soils, sediments and surface waters of the watersheds that drain into and impact the Ocoee River. Acidic conditions and leaching metals have impaired water quality and deforestation has resulted in severe erosion. PCB containing oils have been released to the environment from abandoned transformers. Abandoned and collapsing mine works and other deteriorating facilities and waste piles also pose significant physical hazards. In addition, the lack of a healthy soil structure and the poor quality of riparian and upland ecosystems contribute to poor surface water quality. This has caused the degradation of large portions of the 10,000 acre North Potato Creek and the 3,000 acre Davis Mill Creek Watersheds and 26 miles of the Ocoee River. The two creeks that drain the Copper Basin Mining District site, when left untreated, were releasing over 8,600 pounds of metals and greater than 19,000 pounds of acid into the Ocoee River every day.[57]

In most Superfund cleanups, after determining a need for action, the EPA must list the location on the National Priorities List (NPL). Once the location is listed, the EPA can initiate clean up actions with federal funds based on a structured approach outlined in CERCLA and pursue the potentially responsible parties (PRP) to pay/reimburse the government for the work. More often, however, the EPA attempts to identify a PRP before or concurrent with initiating action. In many PRP-funded cleanups, the project is federalized and managed or monitored by the EPA, often with negotiated input from state and local officials and the PRP. Under such conditions, other applicable laws pertaining to federal agencies must be followed—such as Sections 106 and 110 of the National Historic Preservation Act, which require mitigation of damages to significant cultural resources.

In the Copper Basin, the EPA sought to list the district on the National Priority List, but the issue was fought by Occidental Petroleum—the only remaining district stakeholder with legal liability. Glenn Springs claimed on their website that: "Neither Glenn Springs nor its parent company, Occidental Petroleum Corporation, were ever engaged in any mining activities in the Copper Basin. Occidental Petroleum acquired Cities Service, Inc. after all mining operations had been sold to the Tennessee Chemical Company. Cities Service had acquired the Copper Basin assets from the Tennessee Corporation, which operated the Tennessee Chemical Company."[58] Despite arguing that it was not directly involved in any regional pollution, Occidental had purchased the company (Cities Services) that purchased the company (TCC) that, along with a company it bought (DSCI), was responsible for the majority of the pollution. Because the only other district entities with any potential liability were either out of business (Tennessee Chemical Company), free from prosecution (Boliden Intertrade), or owned by a foreign nation (CITGO), Occidental realized a prolonged fight would entail considerable legal fees.

Further, presuming it would incur even more legal fees associated with an EPA-managed Superfund cleanup and suffer from negative tourism and development impacts for being a district associated with Superfund, Occidental ultimately agreed to assume full financial and managerial responsibility for the cleanup, accept federal monitoring, reimburse the government for its costs, and abide by the general Superfund procedures and protocols outlined in CERCLA. In exchange, the company wanted operational control and the agreement that, as long as the clean-up followed Superfund guidelines and actions were approved by EPA and/or Tennessee Department of Environment and Conservation (TDEC) under a voluntary cleanup and assistance program (VoAP), the operation would remain private.[59]

Although the project remained ostensibly private, Occidental entered into a legally binding memorandum of understanding (MOU) with EPA and TDEC to enforce the cleanup order ensuring state and federal oversight.[60] Similar to fully federalized Superfund cleanups, potentially responsible party (PRP) funds supported studies to determine the extent of damage, proposed and executed cleanup actions, and promoted community involvement in the decision process. A certain portion of the money expended was also allocated directly to community projects as reparation for damages caused and loss of quality of life, as required under Superfund.

Occidental, through its Glenn Springs subsidiary, contributed significantly to district and community projects. It funded annual picnics and site tours, donated to museum programs, provided scholarships to district students attending state universities, and—as it would have under a federally-managed Superfund project—developed a Community Involvement Plan and provided funds to hire technical experts to assist the community's participation.[61] However, unlike federalized projects, Glenn Springs, the EPA, and even the Tennessee State Historic Preservation Office argued when pressed that this cleanup was not legally bound to follow applicable or relevant and appropriate requirements (ARARS) of Superfund or laws that applied to government activities or government property, such as Sections 106 and 110 of NHPA, and neither the federal agency nor the state pursued the issue further. Although this protocol was largely accepted by the stakeholders in Ducktown, an undiscussed question lingered about whether because of the EPA's direct involvement as signatory to the MOU and cleanup monitor, for example, the other appropriate federal laws should have applied.

The EPA divided the region into three main components in the memorandum of understanding. Two were separate watersheds emptying into the third, the Ocoee River (see figure 3.5). The North Potato Creek watershed encompassed most of the original mining lands, the open pits, and the towns of Ducktown and Isabella, including the substantial smelter and acid complex that was largely abandoned by 1987. The Davis Mill Creek watershed included some later mining activity but bisected the Copperhill site that was still active in some capacity through 2008. Although Glenn Springs, as outlined in the MOU, removed a substantial slag deposit adjacent to the Ocoee River in Copperhill and implemented water treatment plants and water diversion controls to contain metals and acid runoff from Copperhill and significantly reduce contamination, most of the cleanup activity focused on the North Potato Creek watershed because Davis Mill Creek still contained active and on-going operating industries governed by other laws.[62]

Ultimately, much of the remediated area in the North Potato Creek watershed will be converted to public park land with hiking and biking trails, and some heritage interpretation, with the Ducktown Basin Museum serving as the park trustee much in the same way it manages the state-owned Burra Burra mine site.[63] By 2008, Glenn Springs had already capped the lead contaminated soils surrounding the Isabella acid plants, and capped and revegetated the former Isabella slag dump and tailings pond. It had fenced

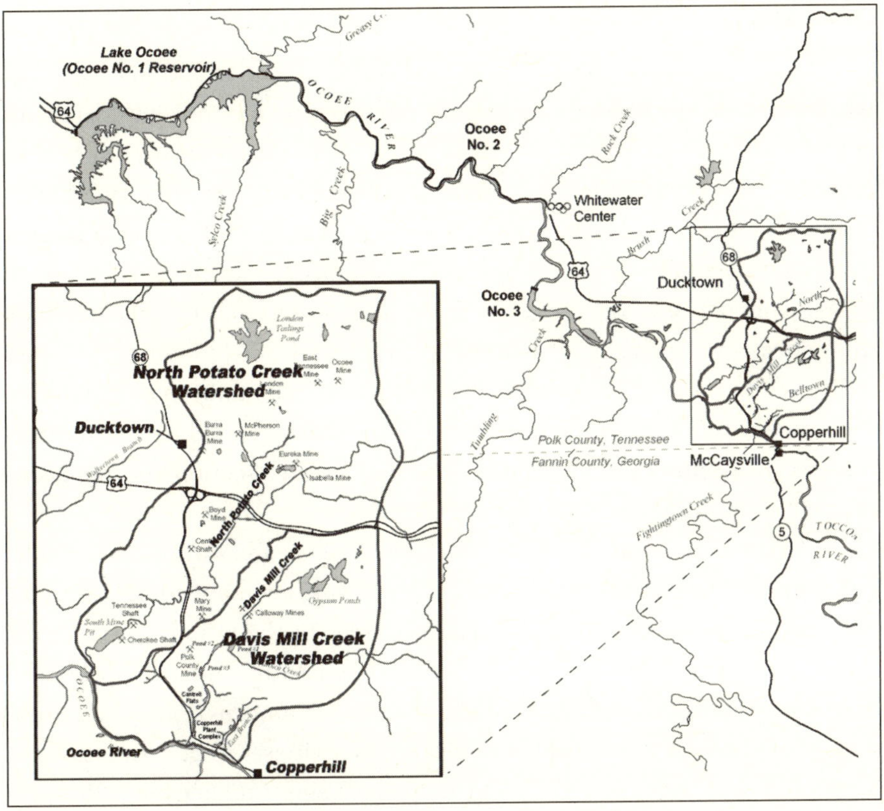

Figure 3.5. Copper basin mining district showing North Potato Creek and Davis Mill Creek watersheds. Courtesy of the EPA.

all of the collapsed mines, removed PCB-contaminated soils, and deposited nonhazardous acidic wastes into the flooded Isabella mine collapse for "sub-aqueous" disposal. Glenn Springs had also built water treatment plants at the South mine pit to process North Potato Creek waters, pumped and treated water from flooded mines, and built an experimental wetland to explore natural water treatment alternatives. All told, according to an EPA website, the water treatment plants in the Copper Basin removed sixteen million pounds of metals and neutralized twenty-eight million pounds of acid by 2009.[64]

Heritage

An important part of any Superfund or Superfund-like cleanup, whether publically funded or privately managed under consent decree, includes public participation. In the original MOU, Glenn Springs agreed to develop and execute plans for public involvement in remediation, education, and future use decisions. The first page of the October 2001 *Community Involvement Plan*, created by the Community Involvement Partnership that included TDEC, EPA, and Glenn Springs, stated that the partnership's mission was to "provide consistent, accurate, and timely information, and to listen and respond to concerns from citizens."[65] In addition to explaining the process and legal framework that Glenn Springs would follow in remediation, the Community Involvement Plan included a section outlining concerns that ultimately resulted in the 2003 *Future Use Plan.*

The *Community Involvement Plan* listed six areas of primary public concern: the cleanup effects on current and future economic developments, the need for preservation of cultural resources in the unique and historic region, the availability of remediated lands for future use, compromised aesthetic land values, potential site-related health and environmental problems, and the cost and duration of cleanup activities.[66] In general, the plan suggested that most residents had reported being satisfied with the proposed remediation efforts and that they had only minor questions about health and safety issues. Other documented opinions included relief that the region was kept off the National Priorities List (NPL), thus avoiding the potentially negative affects on the tourism market, and a strong desire to preserve the unique character of the region and create new recreational opportunities linked to historical and environmental education programs. Still, some long-time residents voiced wistful regret that "the unique moonscape" and deforested red hills could not be preserved. However, many other "residents [accepted and welcomed] the reforestation efforts that . . . restored the tree cover over the last few decades."[67]

Two years after the Community Involvement Plan was completed, Glenn Springs submitted the *Future Use Plan* to "outline the decision-making process for restoring the Lower North Potato Creek Watershed to a condition that is protective of human health, safety, and the environment; and provide . . . for re-development of the site so that it becomes an economic as well as environmental asset to the surrounding communities."[68] This plan only applied to abandoned properties in the district and not to the then-active corridor through Copperhill. A significant portion of the report focused on economic development and tourism, and broke down the demographics of regional visitors and the extent of their monetary expenditures in the area. Most of the plan's new development ideas were geared toward tourism and included new hotels, shopping, and restaurants.[69]

According to the *Future Use Plan*, two primary factors shaped the "sense of place" in the Copper Basin and Cherokee National Forest/Ocoee River areas: nature and outdoor activities, which the reforested lands complemented, and the mining heritage that defined the region's historic identity. This two-part sense of place echoed the long-standing landscape conflict between mining and the natural setting of the district. Whereas the recreational attributes of the region were well established and growing by 2003, the plan admitted that, outside of the museum, "very few historic places remain and none currently provide interpretive opportunities."[70] Of the "few" historic places remaining, the plan cited the structures on the National Register and the three Ocoee powerhouses and dams, but also called for additional heritage preservation, specifically citing the Central head frame and hoist house, Isabella building and acid tank foundations, and parts of the original Old Copper Road. Noting that restored sites fit into a general upswing in nostalgia-based tourism, the plan called for additional thematic developments, such as the historic excursion railroad, addressed specifically to visitors. Summing up the importance of heritage to the region, the *Future Use Plan* stated: "preservation of cultural landscapes such as the barren areas near the Ducktown Basin Museum and the mining structures are important, and displays of how the environment and the people were shaped through the mining activities should be included in the site design."[71] The *Future Use Plan* recognized the community's desire to include heritage in any sort of redevelopment and its understanding that tourism was a—if not the—key factor in future economic growth. It also demonstrated the commitment of Glenn Springs Holdings to the community and its desire to be a more positive corporate citizen—accepting as it did financial responsibility for pollution cleanup while declining any culpability.

Despite Glenn Springs's financial contributions to community programs, the Tennessee Copper district was beset by complications to heritage planning. Heritage preservation in general is not a simple task, and the decisions about what to remember, commemorate, or preserve are contested based on varied perceptions of value. A mill building may be seen by some as an important reminder of what made the region they live in significant, whereas to others it may look like an old dilapidated structure taking away from region's natural beauty. Without a comprehensive cultural resources survey applying the more objective criteria for National Register evaluation, no one knows for sure what exists in the region and which structures or landscapes should receive greater documentation or even conservation attention.

Still, both communities of Ducktown and Copperhill recognized the importance of retaining their historic character if they were going to attract the kind of tourism and economic development they saw as their only viable means of income after their industry had departed. Although the Ducktown Basin Museum and the Copperhill excursion rail terminals both contribute to that end, the district is missing a structured approach to broad heritage-landscape management. Tennessee had (and has) its heritage champions, but Copper Basin's relatively small copper output ultimately did not generate the type of population density that fosters the formation of large, well-organized heritage groups. In 2007, Polk County (TN) only had an estimated population of fifteen thousand, no university or college to draw in new community members, with largely dedicated individuals—as opposed to organizations—promoting heritage projects.[72] The Tennessee Copper Basin, unlike the Michigan and Montana copper districts, had little means of bringing in new people outside of tourists and people seeking vacation homes, few of whom would ultimately become permanent residents.

Tennessee's Copper Basin heritage champions, moreover, were beset by additional complications. Because the cleanup was (and is) privately funded and executed on private land, there was no legal requirement, as interpreted by the government, or regulatory will to enforce cultural resource protection under Sections 106 and 110 of NHPA, and therefore decisions about preservation were left to the groups managing the cleanup. This arrangement did not mean that the cleanup organization and its public monitors did not favor active preservation. Quite to the contrary—they did, but they were not compelled to base decisions on National Park Service–defined levels of significance; rather, decisions were made based on practical solutions without a strong professional heritage group assisting.

Ducktown's mayor, the museum, and local authorities close to the cleanup have all expressed a desire to keep part of the built mining landscape, interpret it, and make it accessible to the public. However, without a legal structure and formal procedure, as outlined in NHPA, or the will to apply National Register criteria, the greater expense of preservation over demolition—and the fact that simple steel or concrete structures are easier to care for than more complicated buildings—tend to dominate decisions. What gets saved is largely based on the cost of preservation and on long-term maintenance rather than on issues of significance and landscape interpretation. Under Sections 106 and 110, important structures too damaged to retain would at a minimum receive some sort of documentation prior to demolition. This contrast is not intended to suggest that Glenn Springs Holdings avoids the issue of preservation, but as the funder of an expensive clean up whose mission is primarily to remediate the environment, it bases its decisions heavily on economic considerations.

The Ducktown Basin Museum faces similar constraints. Ultimately, the remediated land will be managed under a trusteeship by the museum. As the only professional heritage organization directly or indirectly involved in the project, the museum and its small staff can offer opinions about a structure's historical significance, but they have little direct influence over the process of selecting which structures get preserved. The museum is further constrained by the future responsibility of caring for the old structures and the potential liability of dangerous structures. Although they may be historically and culturally significant, mill buildings and smelter foundations ultimately will be very expensive to maintain, especially for a small regional museum with a limited budget.[73]

The selection process for preservation thus becomes very simple in this case. Many decisions about heritage preservation are agreed to by handshake based on windshield surveys and discussions about the demolition and cleanup costs, public accessibility, structural stability, health and safety, and long-term needs. Although these considerations are all appropriate for preservation discussions, without their significance being considered first,

the structures that get preserved are largely limited to concrete bins, masonry retaining walls, some foundations, and simple steel structures, such as the Central head frame.

Although few could argue against the heritage significance and value of the Central head frame and museum buildings, some of the other structures slated for saving have less significance and heritage value in their existing context; for example, the partial remains of the concrete fine ore bins at the London Mill site are being kept. The site of the cleaned bins in their now-grassy field has stunning beauty similar to Stonehenge in the United Kingdom (especially on a sunny summer day), but much of the ore bins' heritage value is lost, because they will be the only historic structure left of a once very large and important milling complex. Without a larger context of surrounding buildings, or at least foundations, the bins are isolated megaliths in a hauntingly beautiful setting, belying their own importance and—without extensive interpretation—incapable of relaying the historic processes that took place there (see figure 3.6).

Figure 3.6. Fine ore bins at the London Mill site before restoration, ca. 2000 (*left*); and in 2008 (*right*). *Left:* courtesy of the EPA. *Right:* by the author.

The decision to save only the fine ore bins reflects the nature of heritage preservation in the Tennessee Copper Basin. The bin's cleaning costs actually exceeded demolition costs, reflecting some compromise with full remediation. But the desire to economically neutralize contaminants associated with chemically hazardous processes at other buildings in considerable disrepair at the site led to the demolition of the remainder of the facility, and thus to the loss of the bin's heritage context.[74] Given a more structured, long-term—rather than narrow, short-term—approach to heritage planning, decisions that included greater considerations of cultural and historical significance might have resulted in a more complete industrial heritage landscape.

Further, the district has no remaining iconic symbol. No single entity in the Copper Basin serves as a replicable and brandable banner image. Although the Central mine head frame comes closest in terms of visibility and recognition by locals, it is not used as consistently as landmarks in other districts. Unfortunately, the one iconic feature commonly associated with

the Basin, the denuded landscape that appeared in myriad photographs and articles for over a century, is all but gone.

Smelter Heritage

The comparative focus of this book is three copper smelting sites in larger mining heritage contexts. If, by definition, smelting includes every process after milling, then roasting, smelting, and refining and relevant by-products must also be included. Therefore, the region affected by smelting in Tennessee includes the entire area denuded by open-heap roasting, and the effort to preserve the small patch behind the museum should be included as smelter heritage along with acid production and the smelters themselves.

The Isabella smelter site, historic home to significant developments in pyritic smelting technology and to the construction of large acid production facilities, is among the most contaminated in the Basin. Near Isabella are the historic roasting yards of the Buena Vista mine, and nearer Ducktown are the McPherson Roast yards. Both of these sites are also considerably contaminated with heavy metals and acids. All three locations were significant to the smelting and mining heritage of the region: the two roast

Figure 3.7. Isabella, ca. 1945 (*left*), and 2008 (*right*). *Left:* courtesy of the Tennessee State Library and Archives. *Right:* by the author.

yards for the smelter smoke they generated that led to the lunar landscape and the Isabella smelter plant for its technology. Much of the Isabella site, however, was dismantled during the 1970s when TCC fully moved metal, chemical, and acid production to Copperhill. It was further damaged by the Isabella mine collapse that is now serving as subaqueous storage for slag waste. The acid plants resulted in considerable lead contamination in the soil—estimated at more than 1,400 tons—and part of the approved remediation plan involved covering the contaminated soil with an impermeable cap to eliminate potential human exposure and leaching.[75] Other potential contaminants at Isabella have been removed, covered, or—in the case of slag—dumped in the pit, leaving little available for historic interpretation. Despite contamination, sound structures, slag piles, foundations, and debris that pose minimal human health hazards are being saved in other districts, even if in a ruined state. Other significant structures and features in the Ducktown district might have been saved, too, if a more structured approach had been attempted (see figure 3.7).

The other significant smelter location at Copperhill has more potential for heritage preservation, but it, too, is in the process of being dismantled by the company operating the site. Because the Copperhill location was active at the time of the memorandum of understanding and not under the jurisdiction of CERCLA or the MOU, the plant was not included in the original cleanup plan and thus was not considered for heritage planning. Because the state of Tennessee is not interested in pursuing heritage considerations in the midst of a complicated nonfederalized cleanup, little will likely be done to preserve any of the significant foundations or remaining structures of what was once the largest acid production plant in the world (see figure 3.8).

Again, this arguement is not meant to imply that Glenn Springs is not interested in contributing to the heritage of the district. Its intentions, efforts, and contributions are well documented. It is simply not bound to consider damages or impacts to properties eligible for the National Register. In general, its decisions are made on the basis of practicality, not historical significance. Ultimately, nearly the entire remediation project area will be

Figure 3.8. Copperhill smelter and acid plant as it looked in April 2008 after years of successive demolition and removal projects. By the author.

converted to a park with walking and biking trails augmented by occasional historic features. But the features selected are being chosen for their durability based not on study or systematic heritage planning, but on handshake agreements following windshield surveys.

In the summary document of the *Future-Use Plan,* the authors assert that the future purpose of the remediation is to "develop revenue generating" recreational assets to attract tourism to the Ducktown/Copperhill region and the state: "Improvements would encourage cultural, historic, and environmental tourism through enhancement of historic mining features and would be important to the region's economy."[76] Specific projects included constructing walking/biking trails, developing interpretive trails and outdoor classrooms to educate the visitor on mining operations, post-mining stewardship practices, and natural history. Other projects included improving educational facilities at the Ducktown Basin Museum, pursuing greater national historic recognition, and establishing redevelopment/enterprise zones in Ducktown and Copperhill to encourage mixed-use office and festive/thematic retail development and to establish excursion rail service from Blue Ridge, Georgia and Etowah, Tennessee.[77]

Although a part of the North Potato Creek site will be converted into an industrial park, the fortunes of the county are aimed squarely on tourism development.[78] With substantial populations in Chattanooga just one hour away, Atlanta two hours away, and Nashville three hours away, and the region's proximity to the Appalachian Mountains and Great Smokey Mountains National Park, the Copper Basin seems like an ideal tourist location, and heritage is at the top of the list of attractions. Although the smelter landscape is largely amalgamated into the broader mining landscape, its historic effects may still be visible despite significant improvements in ground cover. A few small plots of smelter-denuded land throughout the district that did not take to reforestation are exposing stark red soils amid the green canopy, and some of the former roasting yards, now remediated, are slow to revegetate. Interpretive signs will increase the visitor's understanding of the historical processes that shaped the district.

The individuals working to save and interpret some of the historic structures of the Basin should ultimately be commended for their commitment. Much preservation in this country is geared toward tourism; with luck the two efforts will go hand in hand. It is hard to assess how the experiential sensation would be different had a more structured approach to historical significance been applied to determining what parts of the landscape were retained, or how a more structured interpretation plan would combat the stark realization of some inquisitive visitors who, after a period of time hiking and biking in the Copper Basin woods, are suddenly struck that all the trees are exactly the same and all the grasses are exactly the same and all the bushes are exactly the same.

If a single underlying theme permeated the history of the Ducktown District from the first mining attempts through the cleanup and heritage planning, it would be the landscape and its ongoing engagement and conflict with mining, smelting, remediation, and, later, heritage. From the beginning of industrial activity in the area, the landscape and its characteristics defined, and even drove, human interaction more than any other factor outside of economics. The region's remoteness contributed to the first two shutdowns, and its natural features tended toward temperature inversions and limited wind, which contained smelter smoke and contributed to the substantially denuded moonscape. The smoke from the land's sulfur-rich ore killed vegetation and elevated erosion, leading to significant toxin and sediment problems for the Ocoee River, which, as part of the landscape, generated power for businesses and homes and, prior to mining and smelting pollution, provided fish for food. The ore's high sulfur fostered the development of pyritic smelting and created a need for sulfuric acid plants both to protect the land and, through fertilizer production, to enhance the economic capacity of the land. The first remediation treatment, tree planting, was in direct response to the landscape degradation, and later remediation efforts ultimately resulted in a substantial wooded park, whose sheer size and long-term management challenges are affecting heritage decisions.

The other key issues affecting Ducktown District heritage are the privatization of the cleanup effort and the relatively small and isolated community. That the cleanup is private makes many of the active stakeholders' jobs relatively easy when it comes to prioritizing remediation over preservation. The EPA can carry out its mission to protect human health and

Figure 3.9. The central shaft in 2008 towers above State Highway 68 between Ducktown and Copperhill, Tennessee. By the author.

the environment through MOUs and consent decrees not having to be formally distracted by legal considerations for heritage. The cleanup company can largely focus directly on the cleanup and choose how it wants to engage the community. In terms of public relations, this scenario allows the company to respond directly to the specific needs of the broad community and not necessarily become embroiled in controversial heritage decisions. The company is preserving some structures, as well as contributing financially to many other community projects, and most people in the region are happy and satisfied with the company's efforts.

It is difficult to argue that the state of Tennessee is not interested in historic preservation, but with limited budgets and staffing, the State Historic Preservation Office is likely a little relieved that it, too, is not embroiled in assessing significance and enforcing Sections 106 and 110. Certainly the Ducktown Museum would like to retain a greater portion of structures in the remediation area, but it suffers from limited budgets and staff, and already has a fairly sizeable collection of mining structures, so the fewer new buildings the museum has to take responsibility for the stronger its financial stability will be in the long run. Overall, it is difficult to determine how negotiations with the cleanup company would have evolved if the EPA had been forced to ensure much broader heritage planning and enforce payment for greater historic stabilization projects and long-term care.

Moreover, the small population of the Copper Basin meant that organized preservation groups were also small. If the mining district had been more significant nationally in terms of production, then the small preservation contingent might have had a stronger voice for greater heritage conservation. If more of the nationally significant smelting structures, like the Herreshoff pyritic furnaces or part of one of the world's largest acid plant, had survived then preservationists might have had a stronger voice, even if the complexity of remediation and preservation increased significantly. Had the region not been as polluted over one hundred years by metal production and later chemical processing perhaps more outsiders would have been interested in visiting and more structures in the contaminated areas could have been saved. But with only a handful of people promoting heritage tourism as a main goal, and a general satisfaction with the cleanup progress, the community is ultimately getting what it argued for—despite largely top-down heritage decisions.

The approach to heritage in Tennessee is largely an ahistorical and skewed top-down process with EPA and Glenn Springs making the majority of the decisions, as opposed to heritage professionals. This setup, however, is not the only approach. Montana has been largely a bottom-up process with well-organized local groups and members of the community deciding what they want in terms of heritage, then negotiating with the EPA and the PRP to get it or close to it.

CHAPTER 4

MONTANA

> The city wasn't pretty. Most of its builders had gone in for gaudiness. Maybe they had been successful at first. Since then the smelters whose brick stacks stuck up tall against a gloomy mountain to the south had yellow-smoked everything into uniform dinginess. The result was an ugly town of forty thousand people, set in an ugly notch between two ugly mountains that had been all dirtied up by mining. Spread over this was a grimy sky that looked as if it had come out of the smelters' stack.[1]
>
> —Dashiell Hammett, 1929

> ARCO save that stack, touch not a single brick
> Signify the livelihood that made Anaconda tick.
> Still let it stand there stark against the sky,
> Like a somewhat obscene gesture catching every eye.[2]
>
> —Tom Dickson, 2001

November 9, 2008

Driving west on US 90 fifteen miles from Butte, Montana, travelers get their first glimpse—and it is dramatic and huge. It dominates the skyline, but not in the same way that the Empire State Building towers over New York or the Eiffel Tower stands above Paris. The Washoe Smelter stack sits on a geologic plateau seven hundred feet above the valley, the city of Anaconda, and Warm Springs Creek, and in front of a mountain as tall as it. This entire landscape is rooted in the Montana Rocky Mountains, just ten miles from the Continental Divide. Locals enjoy letting you know that the stack, at 585 feet, is as they

have committed to memory, the largest free-standing masonry structure in the world, thirty feet taller than the Washington Monument. But the lone stack sits in a view-shed of equally tall and imposing geologic features.

You know it's big because the Rocky Mountains are big. You know it's big because you can clearly see it from more than ten miles away. Unfortunately, you really can't get a sense of exactly big it is. From even a half-mile away, the stack has still not come into its full height, chiefly because the stack is the only remaining part of what was once one of the largest, most complex, and most advanced copper smelters in the world. Unlike in New York and Paris, no surrounding buildings or structures give it proper scale; in fact, you can't get a real sense of its immense size until you stand right next to it and look up (see figure 4.1).

It is remarkable that the stack was saved from demolition. Certainly one of the most complex, heavily photographed, and—arguably—entertaining parts of many copper smelter reclamation efforts is the demolition of the stack. These structures are often among the tallest free-standing structures in any given county, and their demolition is a very dramatic event that causes as spectacular a shift in the landscape as their erection once did. The 1982 implosion of the Great Falls (MT) smelter stack, for example, was called slow and defiant and was touted as one of the city's most heavily attended events.[3] Following the announcement of plans to demolish the Washoe stack

Figure 4.1. Anaconda smelter stack, from left to right, approximately six miles, one mile, one-quarter mile, and fifty yards away, November 2008. By the author.

just a few months after the Great Falls stack came down, a group of dedicated residents formed a committee to save the Anaconda stack, initially as a memorial to the remarkable efforts of the masons who built it; but, as the movement gained popular support, the soon-to-be-preserved stack came to commemorate the town and smelter workers as well, both still reeling from the shock of the smelter's closure and the deliberate disappearance of other cultural landmarks of their history.

Equally peculiar, but for very different reasons, are the remaining slag wall settling ponds in Butte at the former site of the Butte Reduction Works. This long-lived and innovative early-twentieth-century smelter had immense ponds near its site to store copper-rich tailings. Instead of erecting masonry walls or earth dikes, the company used slag, a waste resource it had on hand. To create the walls, smelter workers erected temporary forms of sheet steel to create boxes roughly 10' (l) x 8' (w) x 4' (h), and poured molten slag into the forms, let the slag set, then moved the walls to the next section and poured again, eventually building up a substantial series of structures—some over fifteen feet high—that impounded an area of approximately one hundred acres. Even more interesting, two of the formed walls created a slag canyon through which Silver Bow Creek still flows. Like the Washoe Smelter, the Butte Reduction Works and all its physical plant have long since been demolished, except for these slag walls. But unlike the smoke stack, whose function is rather intuitive even if its size is not, the slag walls evoke a mysterious, unknown, or little-understood historical process—until one stumbles upon the interpretative sign one-quarter mile off the main street.

Despite the sometimes-confusing nature of seemingly random historic features, heritage pervades the Butte-Anaconda region. A fellow researcher in the Butte-Silver Bow archives liked to say, "History is to Butte, what food is to the French."[4] The ubiquitous symbols of the region's mining history, the gallows frames, are everywhere.[5] The thirteen remaining black steel frames dominate the built environment, and are further emphasized by red cord-lights that outline and illuminate the gallows frames at night. The frames are on license plates, streetlights, newspaper banners, phone books, brochures, and countless local advertisements. They decorate city hall. They are modeled in the county courthouse. They are on business cards and in the City-County Seal and the stylized logos of the convention and visitor's bureau and Mainstreet-Uptown Butte.[6] Several six-foot-tall gallows frame models show up on restaurant lawns, abandoned lots, apartment balconies, and even as a mailbox post in one neighborhood (see figure 4.2).

Heritage symbols fill Anaconda too, but to a lesser degree. The smaller of the sister cities, Anaconda seems to have had less new development and proactive preservation since the smelter closed and, except for the stack, demolished. In addition to being visible from ten miles away, and visible

Figure 4.2. *Left*: Gallows frame of the Original mine in Butte. *Right*: model frame used as a mailbox stand in the Flats. By the author.

in nearly every part of town except the historic cemetery above the county courthouse, reproductions of the imposing masonry structure show up on car dealers' signs, city web sites, and advertisements.[7] "Anaconda Stack" is also the name of the smallest state park in Montana, founded to take ownership of just the stack site and a small viewing area about a mile away. The name of the city of Anaconda's founder, Marcus Daly, not as maligned as in Butte where popular references to him only show up in the name of the Montana Tech lunch shop as a pun on "deli," turns up in Anaconda motel names, coin-laundry names, and in the name of the local historical society, which has had at least four different names since its incorporation several decades ago.[8]

The other and probably second most recognized smelter heritage site in Anaconda is the Jack Nicklaus–designed "Old Works Golf Course" built on the location of Anaconda's pre-Washoe smelter site. Although most of the old works structures were removed shortly after the start of the twentieth century, Nicklaus incorporated existing foundations and waste piles into the course design, and even filled the sand traps with Washoe smelter slag, now called "sland" by locals.[9]

However, not many other popular references to—or celebrations of—waste or toxicity in the region exist, and there's only one currently visible commercial/iconic reference to the wastewater-filled, former open-pit mine, the Berkeley Pit. The *Pit Printers* sits on Front Street in Butte and, at some point in the 1990s, the "Bookley Pit" bookstore operated.[10] But, as in Ducktown, interpretive signs extol the virtues of cleaning up man-made environmental disasters. The Washoe Smelter site, Old Works Golf Course, Butte's gallows frames, and the Reduction Works slag walls are all part of the one of the largest Superfund cleanups in United States history, and at the same time, are part of one of the largest National Historic Landmark Districts in the country. These two competing poles create a rich environment for the preservation of what was once the most productive copper mining district, among whose notable achievements include colossal landscape degradation (see figure 4.3).

Shutdown

The decline of the great Butte-Anaconda district was not linear. The Anaconda Copper Mining Company's (ACM) Montana mines and smelters

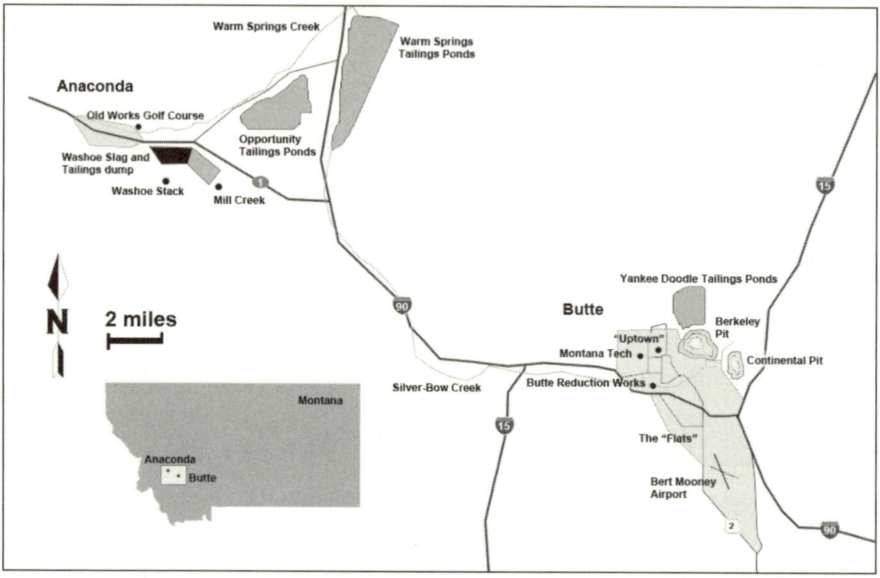

Figure 4.3. Map of Butte-Anaconda showing selected heritage and remediation sites. By the author.

experienced their highest production in 1917, but also saw significant peaks in 1929, 1935, and 1943, each within 82% of 1917.[11] Anaconda's rapid buildup in the early twentieth century included the 1914 acquisition of International Smelting and Refining and its New Jersey, Utah, Arizona, and Indiana smelters.[12] In 1915, Anaconda became the sole corporate entity of Amalgamated Copper holding company and, in 1916, built a zinc plant at Great Falls that, by 1920, was producing over half of the global supply[13] In 1917, Anaconda built a copper wire and rod mill in Great Falls as it began to vertically integrate more fully.[14] By 1919, it owned all but two daily newspapers in the state and successfully fought off—or favorably negotiated—very visible pollution litigation lawsuits from local farmers and the United States government.[15] Further, suspicious union hall bombings and at least one union organizer murder went unprosecuted. In 1922, Anaconda bought American Brass and, in 1929, created the Anaconda Wire and Cable Company.[16] By 1930, Anaconda owned nearly 90% of the copper mining capacity and 100% of the smelting capacity in Montana, held considerable local and state political power and influence, and had expanded operations into Chile and Mexico.

Despite the growth and expansion of the 1920s and the new mining techniques implemented in the 1950s, several historians argue that Montana copper production and the Anaconda Copper Mining Company declined after World War I and never fully recovered.[17] Although it was gradual for several decades, the decline often appeared worse when compared to the rapidly expanding output of Utah and Arizona. Following earlier copper mining investments in Mexico (1906) and Chile (1917), ACM bought a substantial portion of the Guggenheim's Chile holdings in 1923, including the largest single copper reserve in the world at Chuquicamata. As Chile's mines increasingly buoyed profits, ultimately supplying two-thirds of Anaconda's copper and three-quarters of its income by the 1950s, the once singularly dominant Butte-Anaconda copper-producing region became merely a division of a much larger multinational corporation. In the late 1920s, Kennecott's Bingham, Utah, mine began outproducing Butte, and Arizona became the single largest copper-producing state in the country.[18] All this change happened just forty years after Montana had wrestled the designation of largest copper district in the world from Calumet and Hecla and the rest of the Michigan copper mines.

Despite local fears of a corporate pull out, the Anaconda Company continued to invest, mine, and produce copper in Montana, leading to significant World War II output. David Emmons cites combined American and Chilean ACM production figures for 1946, by no means a record year, that included 3.7 million ounces of silver, 30 thousand ounces of gold, 1.47 million pounds molybdenum, 1.55 million pounds of cadmium, 4,883 tons of arsenic, 15.6 million pounds of lead, 1.5 million pounds of zinc, 111,397 tons of manganese, and 742 million pounds of copper, all in the midst of a postwar recession. The Montana component of this production (and the state's relatively small population size) contributed to a 1948 state-wide per capita income 16% higher than the national average, and likely the second or third highest in the nation. Emmons considers the late 1940s to be Anaconda's and Montana's "last hurrah."[19]

On paper, Butte-Anaconda produced at a fairly steady pace from the late 1940s through the mid-1970s with peaks and valleys averaging out to a slight increase over the period.[20] This trend, however, does not reflect significant changes to the mining landscape. In 1947, ACM began the "Greater

Butte Project" to recover lower-grade ores from deposits under the city, because most of the deep veins were so low in copper content that they could no longer be profitably mined. Until this time, Anaconda had recovered all of its copper ore through traditional hard rock stope mining, a selective underground technique in which miners tunnel into ore veins, blast out most of the copper-bearing rock, then proceed down to next level of the mine, repeating the process and sending ore to the concentrators, smelters, and then refineries to produce pure copper. The "Greater Butte Project" initially used block cave mining, an underground form of mass mining that essentially undercuts ore bodies, allowing ores and ceilings to collapse from gravity, then draws off ores and considerable amounts of non ore materials from below. Although it significantly reduced labor and mining costs, this process entailed a substantial increase in rock and ore haulage before concentration.

Despite claims that 3.5 billion pounds of recoverable copper remained underground, plans for a completely interconnected underground system to allow recovery of ores from multiple mines at a single surface head frame, and a nearly twenty-year extension of underground working, most historians consider the "Greater Butte Project" a bust because it was abandoned after only nine years and never produced at the level expected.[21] However the success of the block caving plan was measured, the new underground works were quickly overshadowed by an open-pit mining project that recovered even lower-grade ores using large excavators, dump trucks, and far fewer workers.

The Berkeley Pit began just northeast of uptown Butte in 1955 and quickly ate away at sections of the east side of town. Not only did the new mine absorb several historic underground mine sites and above ground smelter locations on Butte hill, but also eminent domain—or the threat of eminent domain—enabled it to consume the neighborhoods of East Butte, Dublin Gulch, Finn Town, Meaderville, and McQueen.[22] Many of the displaced families and businesses resettled in an area called the Flats, the valley floor south of Butte hill, which had previously housed smelters, ancillary industries, streetcar suburbs, and cemeteries. The Berkeley Pit owed its success largely to the significant quantities of low-grade ores that could be mined with relatively few workers, making the poor ore economically profitable. If the late 1940s were the last hurrah for Butte, the opening of the pit and loss of neighborhoods and mining jobs heralded the last significant decline in the mining district and the beginning of even greater changes to the landscape and identity of the region.[23]

As with all other mining districts, as the copper content of the ore declined, considerably more ore was mined per ton of copper produced. In addition to greater gangue mined per ton of copper, the "Greater Butte Project's" more indiscriminant block-caving methods significantly increased the amount of waste tailings generated by Anaconda, forcing the company to build extensive new settling ponds to keep tailings and dissolved metals from washing into Clark Fork. Operationally, the low-grade copper ore recovered from the pit became too expensive to transport from Butte to the Anaconda concentrating plant, forcing ACM to build a new concentrator in 1964 near the edge of the Berkeley Pit. The erection of this facility effectively closed the concentration department and tailings ponds in the town of Anaconda as ACM opened new ponds and disposed of waste tailings closer to Butte mining operations.[24]

The population of Silver Bow County (including Butte) peaked in 1920 at sixty thousand, but dropped by nearly nineteen thousand, or about one third, by 1970. In 1990 the population would drop to half of its 1920 high. Anaconda's Deer Lodge County, whose population peaked in 1960, would not experience significant declines until 1980, but it, too, would lose nearly half its population by 1990.[25] Amid declining mineral output and population loss, the company divested itself of its newspapers in 1959 and broke its ties with Montana Power in the 1960s, an organization it had once controlled through common board members.

In 1971, a leftist coup forced out the pro-American, pro-Anaconda government of Chile, and new president Salvador Allende nationalized most foreign-owned mines and smelters in the country. This action spelled the beginning of the end for the Anaconda Copper Mining Company. That same year, ACM declared a net loss of $357.3 million and went into receivership.[26] In an effort to trim costs and overhead, the company sold its lumber operations and closed its Anaconda and Great Falls zinc plants. In 1974, it announced it would reduce its workforce by seven hundred to one thousand jobs, and the next year it would lay off an additional fifteen hundred more.

In a further effort to buoy the firm and remain viable, the company began plans to expand the Berkeley Pit into the uptown area of Butte amid a growing public outcry. Despite a rash of unexplained and unprosecuted fires to both uptown buildings and a popular park, and an effort by Butte's mayor to use the progressive federal Model Cities Program to fund the relocation of uptown to the Flats (discussed more fully below), the pit ultimately only got deeper over the next few years and when it expanded it did so away from the town. By 1977, ACM had eliminated a full one-third of its Montana workforce, stopped its last mining operations, closed its Anaconda foundry and, in a last attempt to sustain Butte operations, began mining the East Berkeley Pit, later renamed the Continental Pit to avoid association with the pollution-filled Berkeley Pit just to the west.[27]

These efforts, however, did not save the Anaconda Copper Mining Company. Like Cities Service in the Ducktown district a decade earlier, the Atlantic Richfield Oil Company (ARCO), looking for ways to spend-down some of the considerable profits it had earned during the oil crisis of the early 1970s, purchased the substantially devalued Anaconda Copper Company with a promise to invest millions in the state and to modernize its mines and smelters.[28] ARCO's commitments, however, were short-lived. In 1980, to the shock and disbelief of Anaconda and Great Falls residents, it closed the company's smelter and refinery operations, citing increased pollution abatement costs, and in 1982, it ended operations at the Berkeley Pit and shut off the pumps, allowing the pit to begin filling with toxic water.

Epilogue to Butte Mining

In 1986, the (Dennis) Washington Construction Company of Missoula, Montana purchased the mine properties of the Anaconda Copper Mining Company from ARCO, formed Montana Resources, Inc. (MRI), and began mining the Continental Pit for copper and molybdenum on a significantly reduced scale with nonunion workers—an unusual arrangement that prounion Butte had not experienced since 1934.[29] It also reopened the 1964 concentrator ACM had built at the Berkeley Pit rim and was continuing production as of this writing. More significantly for this work, early on MRI sold several historic mine sites to a new venture called the Montana Mining Properties, which said it intended to restart underground mining in several of the historic shafts.[30] In the process of raising additional capital for its operations, however, Montana Mining put several of the historic gallows head frames up for sale for their scrap value. The thought of losing the last few remaining frames may have been the final event that galvanized the community around preservation. Ultimately, none of the head frames offered for sale was purchased for scrap; all are owned by local government, thus contributing to a truly unique and rich mining heritage landscape.

Washoe Smelter Closure

Unlike the gradual end of ACM, and later ARCO mining in Butte, the end of copper smelting came very abruptly to the city of Anaconda and its workers on September 29, 1980. At the time, labor and local management were in the midst of strike negotiations, but both groups shared the expectation of a settlement and a quick return to work. Just thee months earlier, the Anaconda Minerals Company (AMC)—the name ARCO gave its Montana copper mining division—dedicated a new state of the art visitors center at the smelter's main gate, and the company was in the process of upgrading air pollution controls. Further, just three weeks prior to the closure, AMC was publically considering switching back to coal to fire its Anaconda smelters.[31] In fact, the upper management of the smelter and company-owned railroad were fully expecting to resume operations following the strike and, until the morning of September 29, did not know the smelter would close.

Twenty-five years later, Mel Stokke, the last general manager of the site, said, "The closure was a complete shock, I never thought it would come to pass. Up until the day they told us we were shutting down, I never had a clue." John Greene, general manager of the company-owned Butte, Anaconda, and Pacific Railway, echoed that sentiment: "We were all pretty damn shocked. It just didn't make any sense to us."[32] At that time, even Roger Williams, regional EPA administrator, expressed surprise and deep concern that the company proceeded with the closure, especially after having had recent discussions with ARCO about alternative solutions to abate sulfur emissions.[33]

Despite a mid-1970s effort to bring the Washoe into compliance with state and federal environmental air pollution laws, which included the installation of electric furnaces to melt ores, bag houses to collect dust, and a new acid plant to capture and convert sulfur, the site still fell far short of other regional smelters in terms of emissions control. By the late 1970s, it was the third largest of fourteen western smelters in terms of output, but eleventh in terms of sulfur recovery efficiency, and was cited by the EPA as one of the largest single sources of sulfur dioxide emissions in the country.[34] Attempts to fix the problem aside, the company claimed to be facing increasing difficulty in hitting what it considered a "moving target" of federal and state air pollution regulations.[35] By the late 1970s, ARCO estimated that it needed $300–400 million in upgrades to bring the smelter into full and permanent compliance.[36] Announcing the 1980 closure, James Marvin, ARCO's AMC-division president, said, "The smelter just can't be retrofitted on any reasonable basis and become cost-competitive with modern smelting operations elsewhere in the world." "Since 1972," reporter Shari Meets paraphrased Marvin's statements made the day after the closure announcement, "the smelter operated on a series of variances granted by the state environmental agencies and, as a responsible company, we do not want to, nor do we intend to, continue making applications for variances from the environmental regulations of this state or union."[37] Citing excess smelting capacity in the world in 1980, most notably in Japan, and the high costs of upgrading, ARCO announced the permanent closure of the strike-idled smelter on September 29, 1980. To lessen the blow, ARCO claimed to be considering the construction of a new, state-of-the-art smelter in Montana or maybe an American port city, but neither came to be as just two years later, it ended all copper mining in the district.

Skeptics, recounting conflicting factors in the months leading up to the closure, argued that ARCO was simply interested in minimizing losses and not protecting its workers or caring about its communities. An EPA study requested by displaced workers under the Clean Air Act showed that the smelter could have continued operating under current conditions without any equipment upgrades for eight more years before having to meet new federal air pollution rules and that specific waivers existed for nonferrous smelters under 1977 revisions to the act.[38] Further, the EPA concluded that the cost of permanent pollution controls was grossly overestimated and instead of $300–400 million dollars, should cost ACM between $120–160 million.[39]

More damning claims came from the *Montana Eagle*, a major independent newspaper, in early December 1981. Journalist Bob McCarthy reported that the closure and layoff of fifteen hundred Anaconda and Great Falls employees came just months before the effective start date of the Resource Conservation and Recovery Act (RCRA), which governed the transportation and disposal of toxic/hazardous wastes. The act, however, contained an exemption for companies that ceased operations before November 19, 1980, prompting a former Anaconda City commissioner and several state Health Department officials to claim that this date played a significant role in the order to shut down. The acting plant manager, however, denied this claim despite the fact that he, too, had not been informed of the decision to close until the day of the formal announcement.[40] Whatever factors led to the decision, the McCarthy article demonstrated that the shock and confusion of the closure was still fresh in late 1981, nearly fifteen months later.

To assuage concerns as much as it could, ARCO pledged support to ease the transition and soften the economic blow. Local officials claimed that the closure would result in a twenty-four million dollar annual loss to the Anaconda community in wages and an indirect loss of 1.5 non-AMC jobs for every one AMC job lost. Although another sixteen million dollars would be lost each year in Great Falls, that city's greater population and more diverse economy somewhat cushioned its difficult transition away from copper processing.[41] ARCO claimed that it would help place workers in new jobs, and pay benefits and severance, and pledged five million dollars to the city to attract new development. In 1981, hopes were pinned on a new plastics firm lured with development incentives provided by ARCO, but by 1982 the firm had declared bankruptcy and defaulted on the nine hundred thousand dollar development loan. Overall, ARCO incentives created only thirty new jobs, while fourteen local businesses failed in the two years following the shutdown.[42]

Ultimately little could be done to soften the blow of closure. In the first few years following the shutdown, people left the region looking for work. The state government promised task forces and attempted to lure new industry, but most efforts did little to stem the rising unemployment and growing economic crisis in the town. Worse yet for the once proud town, however, was the growing concern about the toxic residue left behind by the smelter and the growing interest by the U.S. EPA in exploring the site under the Comprehensive Environmental Response, Compensation, and Liability Act (CERCLA), or Superfund, the new federal legislation to remediate abandoned hazardous waste sites. Although ARCO pledged to keep cleaning up problems as they arose, in general, the company felt it was getting close to the end of its responsibility despite the looming Superfund concerns.

Value Transitions

Montana's U.S. Senators had slightly better luck than state and federal programs in combating the declining economy in the Butte-Anaconda region. As early as 1975, the support of Senator Mike Mansfield for magnetohydrodynamics (MHD), a high-tech coal conversion technology, brought a development facility to Butte amid significant copper production declines. Mansfield followed this effort by sponsoring a bill for the creation of the National Center for Appropriate Technology (NCAT), also to be located in Butte. Legislation to create NCAT was signed by President Carter in 1976 and its mission was to apply small-scale and renewable technologies to develop new ways of solving problems common to low-income communities. Appropriate technology ideation, Carrol Pursell wrote, came out of a "broad counterculture movement, a reassertion of doubts about the role of technology in American life, and the burgeoning environmental movement."[43] NCAT, in many ways, stood at the polar extreme ideologically from the Anaconda Copper Mining Company, whose methods were profit driven and anticonservation, and generated considerable waste.

Not that a financially declining ACM, or a financially stable ACM for that matter, would be concerned that a few hundred left-leaning professionals moved into the region, but Senator Mansfield nonetheless successfully headquartered the NCAT in Butte. Though neither it nor the MHD project was intended as an economic replacement for mining, the move was an attempt to change the culture and perception of the region to reflect a more high-tech and diversified future. Both projects contributed hundreds of new high-tech engineering, science, and social science jobs and ultimately led to the first significant influx of outsiders to Butte since the 1950s. These new residents would play an important role in defining the future identity of the region.

With the end of metallurgical operations, the district's identity changed as well, echoing the depth of the region's decline. As with Ducktown, several writers, scientists, and lawyers now made a point of emphasizing the environmental destruction in Montana caused by ACM mining and smelting operations since the nineteenth century. Unlike Ducktown, however, not until copper production ended in the district and the EPA began exploring the area under Federal Clean Air statutes and CERCLA did the negative impact of industrial development, beyond historic smoke-related problems, became apparent to a wider public. The significant waste piles surrounding mining and smelting sites, growing scholarly concern about environmental issues, and the dramatic loss of Mill Creek housing to contamination in 1987 left Montana residents much more aware of—and saddled with—a huge toxic liability that was becoming more visible to the entire nation.[44] Montana historian Michael Malone, writing in 1985, summed up the growing ethos in the state: "The antique Anaconda Smelter shut down in 1980; the landmark Great Falls smokestack was demolished; and the yawning Berkeley Pit now fills slowly with toxic water, perhaps a fitting symbol of the wrenching death of what has historically been Montana's greatest industry."[45] David Emmons, writing in 1994, added: "The discovery that the world no longer needed what generations of Montanans had made had a far more chilling effect on the state's collective psyche than any alleged subservience to ACM."[46]

Punctuating the mounting regional and national perception of defeat, loss, and contamination, a large flock of migrating snow geese landed in the filling Berkeley Pit in November 1995; 342 never left. Although some of the birds died a few days later as they attempted to leave the area, a majority were killed from ingested acids and arsenic poisoning the same day they landed, leaving hundreds of highly visible white carcasses floating on the pit's dark aqueous surface.[47] Although ARCO and the state of Montana vigorously debated the reasons behind the snow goose deaths, the event nonetheless contributed to the deteriorating general perception of the area. In 1993, two years before the snow goose incident, *Travel and Leisure Magazine* called Butte, "the ugliest spot in Montana . . . despite a spirited historic district," and Anaconda, "a sad-sack mining town dominated by a smelter smokestack."[48] By 2009, the region's reputation had hit a new bottom, culminating in a travel designation as one of the world's "must-miss" vacation places.[49]

The designation of "must-miss," however, is far from fair. Since the mid-1980s, a well-organized and dedicated group of Butte, and later Anaconda, professionals and volunteers have been carefully watching the Superfund-mandated cleanup. They have managed the legal framework under CERCLA and NHPA to create not only a safe environment for new economic development, but also a well-managed approach to heritage, resulting in relatively significant structural preservation and regional industrial/environmental interpretation projects, with significant portions funded through Superfund.

Early Reaction to Environmental Degradation

> The mining and processing of gold, silver, zinc, and copper ores generated enormous quantities of solid wastes: those wastes were an inherent and unavoidable part of the enterprise. Clogged streams, mounds of overburden, mine dumps, slag heaps, tailings and slime fields and ponds cluttered the landscape of historic Western mining camps and smelting towns. There were occasions when the miners dumped their wastes directly into adjacent streams and rivers; at other times they stored them, often for decades, behind berms and levee walls. Some of this stored waste became trapped in low-lying marshy areas and some if moved through the swamps and was washed into and down streams and rivers. Occasionally spreading out for hundreds of yards on either side of the waterway; some made its way even further downstream. Because they signaled robust and economically healthy mines, these wastes were not initially seen as problematical. While few had anything to say about mining and mineral processing wastes—whether in place or carried by waters—fewer still had any idea of what to do with them.[50]

Much of the early history of land degradation in the Butte-Anaconda district centers around important wastes: smelter smoke, concentrator tailings, and, to a lesser degree, slag. Although they are not directly part of the smelting process, tailings figured significantly in Superfund decisions in both Montana and Michigan copper districts and had cleanup implications at smelter sites in both areas.

As discussed in Chapter 2, Anaconda's nineteenth-century sulfur dioxide and arsenic-rich smelter smoke killed vegetation and sickened livestock, and the problems only increased as the region's smelters increased production in the late nineteenth and early twentieth centuries. The primary recourses for affected parties, however, were through the courts, suing for property damages or injunctive relief, or by supporting legislative measures to curb pollution. One of the earliest municipal actions in the country against smelter smoke was the 1890 *Butte Smoke Ordinance,* which mandated tall smelter stacks and an end to in-town open-heap roasting.[51] The Deer Lodge farmers' and United States government's lawsuits against

Anaconda's Washoe works in the first decade of the twentieth century were early attempts to formally combat the smoke itself through the legal system. Although these efforts led to the formation of the Anaconda Smelter Smoke Commission and large stacks and Cottrell precipitators to reduce arsenic, none of these efforts significantly reduced smoke and sulfur problems as ACM continued to operate and increase production.

Anaconda, however, used a variety of approaches to minimize its legal risk and continue operating. The company largely settled its early pollution problems through favorable court actions or monetary payments and later sought to minimize potential legal judgments through outright property purchases while searching for technical solutions to the smoke problem to assuage future legal action. In 1902, ACM paid $340,000 to settle claims by farmers and ranchers and, in 1903, built the first large stack and flue system at the Washoe to reduce local smoke emissions. During this period, it also conducted experiments to determine the affect of smelter smoke, operated ranches and farms under the smoke plume largely to challenge would-be litigants, and acquired smoke (and tailings) easements from other property owners.[52] It also purchased some affected farmland outright to avoid civil penalties and began acquiring land for tailings storage. Early twentieth-century land purchases increased Anaconda Smelter holdings in Deer Lodge County from 5,945 acres in 1906 to 7,014 in 1917, and Anaconda's other Deer Lodge holdings from 3,000 acres in 1906 to 19,139 in 1917.[53]

Following the Department of Justice's (DOJ) 1910 lawsuit against Anaconda, DOJ and ACM established the Smoke Commission in 1911 to monitor pollution and recommend alternative practices to abate future problems. As long as Anaconda consented to the commission's recommendations in good faith, the DOJ would not pursue the lawsuit in court. The commission ultimately had a hand in the decision to build the 585' Anaconda Stack and to install Cottrell arsenic precipitators as mentioned above.[54] By 1920, historian Fredrick Quivik surmises, the Justice Department was ultimately satisfied that the Cottrell precipitators recovered 80% of the arsenic, that the new stack sufficiently disbursed smoke, and that no profitable means for complete sulfur removal was available in Montana. Further, Anaconda was satisfied that the processes and equipment recommended by the commission ultimately increased profits by recovering saleable by-products without disrupting smelting operations and, because the Government lawsuit was not formally prosecuted in court, ACM did not have to pay damages or costs.[55]

Although the Justice Department was largely satisfied with the efforts of Anaconda, the United States Forest Service (USFS) was not. In 1922, it identified 150,000 acres of forest worth $1.3 million in lumber, land, and grazing potential damaged by smelter smoke. Instead of taking the corporation to court, the USFS proposed a land swap of equal value properties, and Anaconda consented. By 1932, in several separate agreements, the United States and Anaconda exchanged 84,000 damaged acres for 95,000 valuable acres in southwest Montana. Additional land swaps were negotiated and some later executed with the Forest Service, but the Justice Department, feeling there was nothing left for it to do regarding its 1910 legal action, finally abandoned the lawsuit in 1933.[56]

The acquisition of Deer Lodge and Silver Bow county lands had one major culminating effect. By 1933, when the Justice Department dropped its lawsuit, ACM had largely settled all of its early pollution issues by buying or trading the agriculture, ranch, and forest lands its smelting operations had affected, and, either buying acreage or purchasing easements on all the other property it needed to dump its waste. By the mid-1930s Anaconda operated with a pollution buffer surrounding its operations that would enable it to operate without interference until the passage of federal regulations to control the transportation and disposal of hazardous wastes starting in the 1970s (see figure 4.4).

Origins of the Butte-Anaconda Heritage Movement

> She's battered, and she's beaten,
> She's ugly and she's torn,
> She's battle-scarred and gutted,
> From the Copper she has borne.
>
> Still she is majestic,
> As she sits upon her throne
> A mile high and beautiful,
> To those who call her home.
>
> —George T. Grosse (1957, Quoted in Chavez, 1982)

Reading through the secondary literature on the Butte-Anaconda district, one comes across many poems and anecdotes reflecting the view that, although their region is regarded as dirty and dangerous, and the ACM corrupt, local residents were still proud of their communities and their work. Baum and Knox quoted a retired miner who said, "Them times was tough, but we had no worries about that. Anaconda owned this town, and Anaconda put food on the table."[57] However, as underground mining slowed considerably in the 1950s and the Berkeley Pit began consuming

Figure 4.4. "'E' Smoke gets in their eyes." An end to pollution litigation did not mean an end to smoke problems or even a denial of its existence. This cartoon appeared in *The Mining Journal* in 1943 reflecting age-old ideas that smoke was a sign of a healthy, robust, and, in this case, patriotic industry.

neighborhoods, the community members remaining after redundant parts of the population left to find work elsewhere began to think about the former days when its mines and smelters were well known for being some of the largest producers on earth. Although the decline of mining in Butte-Anaconda may have initiated heritage directly, the drive to keep mining and to expand the Berkeley Pit had a much more powerful influence on the first attempts to document and preserve the region.

In much the same manner as those in the Michigan District in the 1950s and in the Ducktown district in the 1970s, residents of Butte-Anaconda opened a museum at an abandoned mine site in the 1960s and worked to get its significant structures listed in the National Register of Historic Places following an earlier National Historic Landmark designation for parts of the mining city in 1961. The Butte Exchange Club started the museum in 1963, and by 1964 had found a permanent home at the Orphan Girl Mine, donated by the Anaconda Copper Mining Company. Between its official opening in 1965 and its one millionth visitor in the 1980s, the World Mining Museum amassed an eclectic assemblage of mining artifacts, many donated by workers or by ACM itself, and added a mining hall of fame, a recreated nineteenth-century Wild West mining town, and above-ground open-air exhibits on mining. By 2008, it had added an underground tour, a memorial garden to fallen mine, smelter, and railroad workers, and a significant photo archive.[58]

Although the decline of mining in 1950s and 1960s caused locals to reflect on their past and open a museum, the Berkeley Pit—and later the threats of pit expansion—caused a much more dire threat to the district than the decline of mining. Where the Michigan and Tennessee districts built museums and lost structures slowly to neglect during the decline and abandonment phases, neither district faced the large-scale destruction of cultural resources until the Superfund period. Butte encountered this situation decades earlier. The creation of the Berkeley Pit destroyed neighborhoods, industrial areas, recreation areas, and churches, and the threat of expansion caused citizens to react by starting to preserve their heritage.

The initial construction of the pit in the early 1950s, and its growth in the ensuing years, had clear ramifications for the identity of the district. According to the Butte-Silver Bow Archivist, the early growth of the pit caused Butte residents to feel trepidation about its expansion, and one local resident, the owner of one of the original "Copper King" mansions, initiated contact with the National Park Service to get portions of Butte listed as a national historic resource.[59] This early effort, although affording little actual protection against pit expansion under eminent domain, resulted in a 1958 survey leading to a National Historic Landmark (NHL) designation for parts of Butte in 1961.[60] The initial district covered several uptown blocks and, with later assessments following the NHPA in 1966, would include much of the Butte hill.[61] These earlier efforts would be expanded in 1985 to include a new section formally documenting industrial buildings and mining sites, and the boundaries would be significantly enlarged in 2005 to include Anaconda.[62]

In the 1970s, two more events struck at the heart of an economically depressed Butte. First was a series of fires, some resulting from faulty wiring in old buildings, but more troubling were the scores of arson and fires of unknown origin. Between 1972 and 1975 more than twenty major buildings in the uptown historic district burned.[63] Although several of the arsons went unprosecuted, many locals suspected that at least some of the fires were authorized by ACM preceding the planned Berkeley Pit expansion. Locals recounted stories about friends of friends who earned five hundred dollars for burning buildings in the 1970s, and others claimed that the company had used minors as its primary arsonists to allow them to avoid "adult" jail if caught.[64] At the same time that fires were burning uptown, new commercial developments were being constructed on the Flats, including a shopping mall, new airport terminal, restaurants, and housing tracks. By 1975 more than thirty businesses had fled the central business district (CBD) for new retail developments in the Flats.[65] Further, the Interstate 90 corridor had significantly altered the traffic patterns of the city, allowing drivers to bypass the older uptown stores and easily access the new Flats commercial district.[66]

The city of Butte was shifting away from its historic business district uptown to the sprawling flats south of the hill. Although suburbanization was a common feature in most extraurban growth areas after WWII, Butte had one additional uncommon external pressure on its older surviving neighborhoods—the pit. By the 1970s, most community leaders seemed resigned to the eventual loss of the uptown area. A map accompanying a revision to the 1960s National Historic Landmark nomination form demonstrated the extent of the expected pit expansion (see figure 4.5).

With the decline in mining, an operating museum, and a National Historic Landmark district, the heritage movement had a fairly typical start, very similar to that of the Ducktown district and many other historic industrial districts in the country. However, the heritage movement in Butte would grow significantly as a result of three more events: the "Butte Forward" plan to relocate the CBD, a significant reduction in funding for the NCAT, and, like the other districts, the advent of Superfund.

Prompted by the ACM's December 1975 letter detailing the company's need to expand the pit, which everyone in town had been expecting for years, Butte's mayor proposed the "Butte Forward" plan initiated under the federal Model Cities Program. This program granted funds for infrastructure repairs, administrative training, and reorganizational studies under the broad rubric of "urban renewal." Under the plan announced in 1976, city officials created a public-private development group called Butte Forward, Inc. to plan for and execute the relocation of the central business district. Later that year, the group submitted its fifty million dollar proposal to the city, which included expected contributions of eleven million dollars from ACM and thirty-nine million dollars from the U.S. government. Despite a petition effort by uptown merchants calling for a public vote on the matter—and some debate on the actual location of the new business district—the city council voted eight to five to move ahead as planned, but, because of a procedural matter, tabled the issue and planned to officially reconsider the measure two weeks later.[67]

To the consternation of the mayor and ACM, the council rejected the plan on the second vote, in a move that historian Brian Shovers suggested was possibly the most important decision in the city of Butte's one hundred–year history. Shovers writes, "In defying relocation [and rejecting the century-long dominance of ACM] Butte not only reaffirmed its historic identity, a persona forged by the rigors and spawned by the culture of underground mining . . . but had drawn the line on how much it might sacrifice in the name of a paycheck." Almost immediately, Montana Power, which held several seats on the Butte Forward, Inc, board, began purchasing historic buildings and refurbishing them for tax credits, while other organizations—realizing that the city made a significant decision to embrace its historic identity—began shifting priorities.[68]

Although many still did not fully support heritage as an economic model, several efforts began to funnel money into restoration projects. In

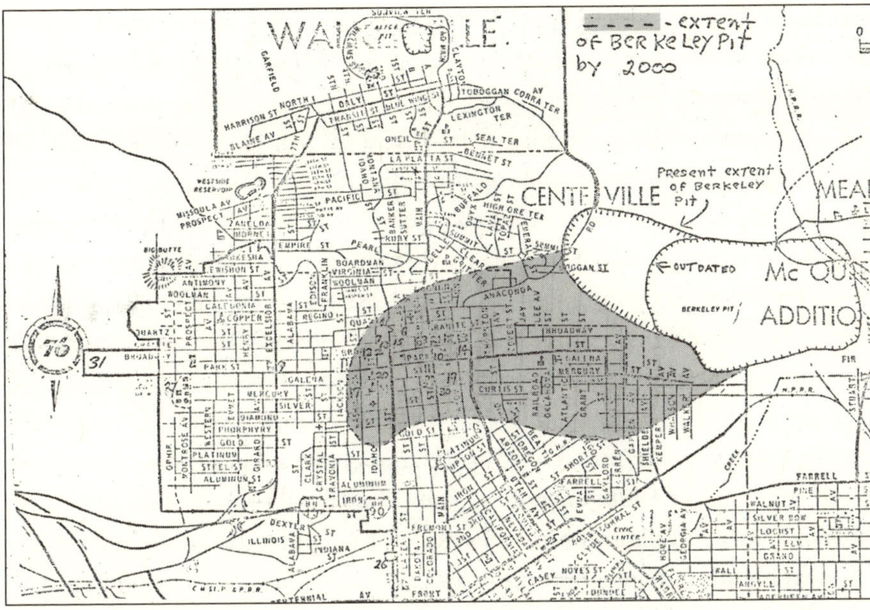

Figure 4.5. Extent of the expected Berkeley Pit expansion from 1970s to 2000 included the primary central business district, courthouse, and several churches, restaurants, bars, and schools. This particular map was used to argue for the creation of the National Historic Landmark district. Ultimately, however, the pit did not expand into the central business district and remained at the size indicated by the "present extent of the Berkeley Pit." Courtesy of the U.S. National Park Service.

1979 and 1980, the Historic American Engineering Record (HAER), a department of the National Park Service (NPS), was invited to Butte under sponsorship of the local government to formally document inactive industrial features of the landscape.[69] At the time, HAER was a part of the Heritage Conservation and Recreation Service of the NPS. Although the documentation of historic industrial sites was a paramount component of its mission, the HAER team was also charged with assisting in the redevelopment of blighted areas and, in this instance, "analyzing and developing a revitalization strategy for Butte's uptown historic district."[70] In addition to photographs, drawings, and a written history, the project proposed the creation of a revitalization agency, funded by tax increment funding, to manage redevelopment in Butte's historic district. Although Butte and parts of Anaconda would be documented by HAER during the 1980s, the initial project would bypass the Washoe Smelter largely because it was a still-active site when the initial documentation began, then demolished before the later documentation efforts started.

Around the same time as the HAER project began, a local resident noticed large dump trucks pulling up to the former ACM office in Butte to dispose of company records. After stopping the destruction of records, he ultimately formed a group that worked with the county executive to establish the Butte-Silver Bow Public Archives in 1981 with considerable volunteer hours and, later, significant public-approved funding.[71] The archives grew to encompass extensive regional corporate and labor records.

In 1977, Butte consolidated with Silver-Bow County into a single governmental entity. The new consolidated government followed the 1980 HAER recommendations and created an Urban Revitalization Agency (URA). URAs are basically planning and development organizations designed to use tax increment financing to promote infrastructure and restoration improvement projects within a specified district, which in this case was the Uptown Butte area.[72] The additional tax collected on the increased value of restored or rehabilitated buildings was then pooled in a fund managed by the URA and reallocated for new restoration and development projects in the historic district that generated further tax revenues.[73]

The growth of heritage interests in the region was further enhanced by setbacks to programs that residents had hoped would offset the loss of mining jobs. In 1980, Ronald Reagan was elected president and set out to reduce the size of the federal government.[74] Reagan slashed budgets and social programs that did not generate a strong economic impact and attacked those who, according to Pursell, refused "to embrace technological vanguardism as the finest expression of national virility."[75] Two affected programs were the NCAT office and MHD labs in Butte. Both organizations had brought highly educated and, arguably, activist people to the tough but fading blue-collar mining town of Butte to work on high tech and politically progressive programs. When budget cuts hit, the MHD lost several hundred people, and NCAT went from over one hundred and twenty people to about ten in a very short time.[76] Many of those affected stayed in Butte to continue their work and research and would become involved in heritage preservation. Several people, for example, formed a corporation reflecting the NCAT mission, Renewable Technologies, Inc (RTI), a mixed-mission organization that allowed individuals to pursue personal projects, including supporting preservation agendas within the larger Superfund project.[77]

Between the 1960s and 1980s, heritage professionals later reflected, Butte and Anaconda's long-term mining residents still hadn't fully embraced heritage as either an expression of identity or potential market for tourists. The long-term residents were often still numb from the closure of the mines and smelters and psychologically shaken by the spate of fires and uncertainty of their town's physical presence.[78] They felt that the heritage movement was driven in many ways by outsiders who, like ARCO, valued the built and cultural landscape differently because it didn't hold the same meanings for them. Further, many of the outsiders came after the demise of the Anaconda Copper Mining Company, the loss of neighborhoods with the construction of the Berkeley pit, the rash of unexplained fires, and the "Butte Forward" plan.

City of Anaconda residents lagged even further behind those of Butte in industrial heritage conservation planning despite a significant effort to preserve the Washoe stack and the creation of a local historical society. Whether it was Anaconda's smaller population, the "still-festering" shock of the smelter closure, or the lack of any new economic development, Anacondans in general seemed to give industrial heritage a lower priority until the Butte movement started thinking of—and planning for—an expanded historic copper district in the late 1990s and early 2000s, which included Anaconda smelter, foundry, and transportation sites.[79] In fact, even the preservation of foundations and waste piles at the old-works golf course came at the insistence of Jack Nicklaus rather than from local development officials.[80] Ultimately, not until the work of the new outsiders began to reshape the district and significant infusions of Superfund-generated money became available for remediation projects did the perception of the "ugly town, set in an ugly notch between two ugly mountains" begin to change.

Superfund

A century of mining and smelting in the Butte-Anaconda district resulted in several significant modifications to the landscape. Not only did it create thousands miles of underground mine shafts, drifts, stopes, crosscuts, and tramways, but it also dug one of the deepest open pits, built the largest smelter complex in the world in the 1890s (and then again in the 1900s), erected the tallest masonry structure in the world, and produced over thirty billion pounds of metals worth over twenty-two billion dollars.[81] However, in mining and processing the Butte copper lode and extracting significant profits from its sale, the industry was ultimately responsible for the landscape changes that resulted from dumping and discharging its wastes, most of it legally, into the local environment. The significant scale of arsenic and sulfur-laden smelter smoke releases, acidic dusts, metal-rich waters emanating from tailings deposits, mine drainage, the Berkeley Pit, and significant tailings and slag dumps led, not surprisingly, to serious and extensive problems. The following passage comes from the EPA websites for Silver Bow Creek/Butte Area Superfund Site:

> More that 100 years of mining have resulted in the development of over 500 underground mines (with roughly 3,000 miles of underground workings) and 4 open pit mines including the Berkeley Pit with its ancillary tailings ponds, waste dumps, and acid leach pads. Operation of silver mills and copper and zinc concentrators/ smelters in Butte resulted in the production of a variety of mill and smelter wastes including particulates (aerial emissions) and tailings. The long period of mining in Butte left the landscape littered with un-vegetated or sparsely vegetated mine wastes, often containing hazardous concentrations of metals and arsenic. These wastes represent significant sources of environmental contamination to Silver Bow Creek and posed human health and risks to the environment.
>
> Ground water, surface water and soils are contaminated with arsenic and other heavy metals, including copper, zinc, cadmium and lead. Silver Bow Creek and the Clark Fork River contain metals from the cities of Butte to Milltown. The tailings, dispersed along the creek and river, severely limit aquatic life forms and have caused fish kills in the river. Potential health threats include direct contact with and ingestion of contaminated soil, surface water, ground water or inhaling contaminated air.[82]

Similarly, for the Anaconda Smelter Superfund site:

> The processing facilities at the site were developed to remove copper from ore mined in Butte from about 1884 through 1980. Milling and smelting produced wastes with high concentrations of arsenic, as well as copper, cadmium, lead and zinc. These contaminants pose potential risks to human health, to life in nearby streams, and to plants and animals in adjacent lands over some 300 square miles. In addition to the millions of cubic yards of tailings, furnace slag, flue dust, and square miles of soil contaminated by airborne wastes, millions of gallons of ground water have been polluted from wastes and soils. Arsenic is the primary COC and drives the remediation.[83]

The scale of mining waste emanating from the Butte-Anaconda region ultimately led to three separate Superfund listings on the National Priorities List, each with multiple individual operable units to coordinate the massive scale of the cleanup.[84] (See table 4.1) The *Silver Bow Creek/Butte Area,* officially listed on September 30, 1983, then expanded on July 22, 1987, had thirteen individual operable units encompassing Butte mine and smelting sites in the Silver Bow Creek watershed, metals-contaminated residential soils, contaminated mine, pit, and ground water, and the extensive tailings strewn along the length of Silver Bow Creek from Butte down to, and including, the Warm Springs settling ponds outside of Anaconda. The *Anaconda Company Smelter Site,* listed on September 30, 1983, included ten operable units in the Warm Springs Creek watershed relating to three smelting locations and ancillary industries, contaminated soils and house dusts, and the Opportunity tailing ponds just outside Anaconda. In each of these regions, the chief contaminants were heavy metals that existed in smelter flue dust, including significant amounts of arsenic and copper that settled on plants, soils, and inside people's houses, and water contamination with dissolved acids and metals, including sulfuric acids, arsenic, cadmium, copper, lead, and zinc that flowed from groundwater into wells and discharged into waterways.

The third Superfund site relating to mining and smelting was the *Milltown Reservoir/Clark Fork River* area, which included 120 miles of the

TABLE 4.1. List of Superfund Operable Units in Butte and Anaconda

Silver Bow Creek/Butte Operable Units and clean up action[1]			Anaconda Co. Smelter Site Operable Units and clean up action[2]		
1	Streamside Tailings	Remedial	1	Anaconda Smelter	Removal
2	Area One		2	Anaconda Yards	Removal
3	Butte Mine Flooding	Remedial	3	Arbiter	Removal
4	Warm Springs Pond	Remedial	4	Beryllium	Removal
5	Reduction Works Tailings	Removal	5	Old Works	Removal
6	Travona Mine	Removal	6	Mill Creek	Removal
7	Rocker	Remedial	7	Flue Dust	Remedial
8	Butte Priority Soils	Remedial	8	Community Soils	Remedial
9	Active Mining and Milling	Remedial	9	Old Works/East Anaconda Development	Remedial
10	Residential Soils	Removal	10	Anaconda Regional Water, Waste, Soil	Remedial
11	Lower Area One	Removal			
12	Warm Springs Ponds Inactive	Remedial			
13	West Side Soils	Remedial			

Removal actions are intended for the immediate removal of direct threats to human health and the environment while *remedial actions* are intended to be long-term, permanent solutions to contamination.

1. See, Region 8 U.S. EPA, ed., "Second Five-Year Review Report for Silver Bow Creek/Butte Area Superfund Site" (Helena, MT: CDM, 2005).

2. See, Region 8 U.S. EPA, ed., "Third Five-Year Review Report for Anaconda Company Smelter Site Anaconda, Deer Lodge County, Montana," (Helena, MT: U.S. EPA, Region 8, 2005).

Clark Fork River from Anaconda down to and including the Milltown Reservoir outside Missoula, Montana.

This site was broken into just two operable units, the Milltown Reservoir, which entailed 2.2 million cubic yards of impounded contaminated tailings, and the Clark Fork River's entire length, which was strewn with tailings that did not wash all the way down to Milltown. Here, too, the primary concern was heavy metals that precipitated and settled out of water and contaminated Clark Fork riparian and impacted Milltown soils. In both operable units, the primary solution was to remove contaminated soils and tailings and to deposit them in the still-open Opportunity Ponds in Anaconda.

Unlike Occidental Petroleum in Ducktown, ARCO accepted the initial NPL listing but, like Occidental, resisted active participation for a decade, claiming that the oil company hadn't polluted the region directly.[85] Further, ARCO contended, as the state of Montana and the EPA sought additional penalties, that the state and federal governments were also potentially responsible parties because they both encouraged mining for regional and national gain and did not create or enforce stricter pollution legislation earlier. ARCO eventually hired its own historians to prove its case in court.[86] These early arguments, however, were largely moot by 1983, when ARCO consented to pay for the initial remediation projects.

Nonetheless, court challenges continued through the 1990s and into the 2000s, especially after both the state and federal governments sought to recover their response and remediation costs and impose civil fines for reparations and community projects. ARCO responded to the new charges and filed counterclaims against both the U.S. government and the state of Montana to recover *its* response costs and contributions, and to receive "contractual indemnity, equitable indemnification, and declaratory relief," citing the government's conspiracy to develop and promote mining in the western United States.[87] Ultimately, however, with a series of negotiated consent decrees from 1998 to 2008, ARCO agreed to pay additional remediation costs and damage claims, albeit less than originally sought by the lawsuits, and to cover governmental cost and oversight expenses.[88]

Needless to say, the entire remediation project emanating from mining and smelting in the Butte-Anaconda region led to significant Superfund cleanup costs and, in terms of geography, became the largest effort executed in the country at the time. By 2008, ARCO had agreed to pay the state of Montana $230 million for natural resource damages to Silver Bow Creek,[89] $168 million for damages associated with the Clark Fork River and other sites, and nearly $14 million to the U.S. government for EPA costs, and to settle damage claims made by the National Park Service and Bureau of Land Management.[90] Additional costs included an estimated $316 million to remove sediments from the Milltown Reservoir and even more for other cleanups, community development projects, EPA and state fees, and long-term waste monitoring and remediation projects for various sites in the districts.[91] As early as 2002, ARCO expected its total bill for clean up in Montana would exceed one billion dollars.[92]

Although it vigorously pursued ARCO under CERCLA to—as its mission states—protect human health and the environment, the Environmental Protection Agency was less concerned with ancillary requirements of CERCLA, such as abiding by the *Applicable or Relevant and Appropriate Requirements* (ARARs) as required under section 121(d). Almost from the beginning, the EPA eschewed its responsibilities to fully consider impacts to cultural resources eligible for inclusion on the National Register, as required under Sections 106 and 110 of the NHPA. According to an official directly affiliated with cleanup monitoring for the U.S. government, the EPA met only the minimum necessary requirements of Sections 106 and 110 *as it interpreted them* and, in 1987, simply took photographs of buildings it thought might be impacted.[93] Jerrold Marcus, writing in the *Engineering and Mining Journal,* states the EPA only had two strategic goals: remediate the site and allocate as much of the cost as possible to the potentially responsible parties. Further, EPA officials understood that as a lead agency, it would be "severely criticized" for its decisions by other stakeholders no matter what it did.[94] Heritage professionals fell into this latter group.

Mark Reavis, former Butte-Silver Bow County preservation officer, recounted that he had to fight the EPA and ARCO to get them to comply with NHPA and to retain the region's significant historic character.[95] He later argued that "Odd as it sounds, those dumps are historic resources. The preservation community here is worried that we're going to lose, bury, and cover-up all signs of mining. Butte should be a monument to a social decision: the quest for minerals. I'm trying to preserve. They're trying to clean up."[96] Similarly, Fredric Quivik, who came to Butte to work for NCAT in 1977, wrote in 2001, "The EPA [in Montana] . . . often pursued its mandate to remediate hazardous materials with a single-mindedness that has not recognized other national priorities, such as the preservation and interpretation of important cultural resources. . . . Local [EPA] administrators seemed to see the historic copper industry as the enemy, and they saw the only design alternative for remediation to be the obliteration [of] any historic features that happened to be in the way of the cleanup and to cover those areas with newly planted grass."[97] The Anaconda-Deer Lodge Planner acknowledged that Superfund benefited the health of the community but noted that it could be detrimental to historic preservation. Further, this authority cited instances in which the EPA manipulated Sections 106 and 110, claiming that sites it didn't want to remediate were "historic," while other sites it wanted to clean were not.[98]

The Butte-Anaconda region, with the largest Superfund sites and one of the largest National Historic Landmark districts, highlights the inherent conflict between remediation and preservation. Unlike Ducktown, where the heritage decisions were largely left in the hands of the remediaters, the Butte-Anaconda district, with combative new residents and a strong Montana SHPO and heritage community, ultimately established procedures and negotiated heritage preservation with both the key national heritage organizations and the EPA and ARCO, although during the planning process many sites, like the Washoe smelter, were lost.

Superfund-Generated Heritage

In Butte-Anaconda, the clout of heritage preservation advocates grew slowly between the mid-1980s and 2005, and included the creation of county-level preservation officers to oversee heritage management plans. (See table 4.2 for a summary of dates.) Although Superfund/ARCO funded or contributed to many of these plans, the genesis of these efforts was often local reactions to the potential loss of cultural fabric from cleanup activity that had mostly been driven by the new, proactive outsiders captivated by the region's rich heritage, but not affected as much by its decline.

According to the Butte-Silver Bow Planner, the URA gave birth to the first comprehensive preservation ordinance in the region.[99] Although the National Historic Landmarks (NHL) office established an architectural district in 1961, and Butte established physical boundaries for tax increment–funded redevelopment in 1980, the URA created a new historic overlay zoning district in 1985—twice the size of the earlier urban revitalization area—to promote the preservation of historic sites, structures, and buildings by "addressing preservation issues at the local level and integrating them into the planning and decision making process." The ordinances' specific goals were

TABLE 4.2. Timeline of Preservation and *Remediation* Events in Montana

Dates	Event	Location
1961	First Butte district NHL listing	Butte
1965	Mining Museum Opens	Butte
1967	Underground mining ends	Butte
1971	Chile nationalizes mines	Butte
1975	MHD lab opens in Butte	Butte
1976	NCAT office opens in Butte	Butte
1976	Threat of Butte CBD relocation	Butte
1977	ARCO buys ACM	Butte
1979	HAER documentation	Butte
1980	URA begins	Butte
1980	Washoe Smelter closes	Anaconda
1981	Butte Silver Bow Archives begins	Butte
1981	MHD and NCAT budgets slashed	Butte
1982–86	Washoe Smelter Demolition	Anaconda
1983	*Superfund National Priorities List*	Butte-Anaconda
1985	First Butte NHL revision	Butte
1985	Washoe Smelter Stack saved	Anaconda
1985	Butte-Anac. Historic Park Master Plan	Butte-Anaconda
1988	*General Work Plan for Anaconda cleanup*	Anaconda
1990	*RI/FS for Anaconda*	Anaconda
1993	Regional Historic Park Plan	Butte-Anaconda
1997	*Old Works Golf Course opens*	Anaconda
2005	Second NHL expansion, includes Anac.	Butte-Anaconda
2008	Butte Reduction Works slag walls saved	Butte

to "enhance the visual character . . . encourage maintenance of the present housing stock . . . promote [cultural] tourism . . . enhance property values . . . foster public appreciation of and civic pride in the beauty of the community and accomplishments of the past . . . and to safeguard the heritage of the community."[100]

More significantly, the new ordinance specifically included the existing thirteen head frames and defined historic buildings to include "appurtenant mining structures."[101] It also created a historic preservation commission to administer the preservation ordinance, a historic preservation officer to enforce codes that required a review, a permit process to move, remove, or demolish historic structures, and created appeals, enforcement, and penalty policies for violations. Before defining the boundaries of the new district, however, the new ordinance required a new survey to identify structures.

The years from 1981 to 1986 witnessed several surveys, documentation efforts, and planning processes to catalog historically significant structures in Butte, expand the original National Historic Landmark district to include industrial sites, consider sites in Anaconda, and begin planning for an eventual National Historical Park dedicated to American copper. Following the Historic American Engineering Record documentation project, the Butte Historical Society began a five-year project to survey all historic resources in the historic district.[102] In 1985, the Historical Society, with additional financial support from the state of Montana, community development grants, and a host of additional sources, published an RTI planning study/survey of the architectural and industrial heritage of the Butte-Anaconda corridor that ultimately created a planning basis for preservation. Not only did this study put the entire mining/smelting district into a single historic context for the first time, but it also served to provide reference points for future restoration projects and long-term planning efforts.[103]

Two significant planning documents followed this effort. First, with the realization that so much industrial heritage still existed, the original 1961 National Historic Landmark District was expanded to include mining sites on Butte hill. More importantly, the study led to the first comprehensive heritage plan for the combined mining and smelting districts. The *Butte-Anaconda Historical Park System Master Plan* was written by RTI also in 1985 and published by the Butte Historical Society with funds from Anaconda Minerals Company, the National Trust for Historic Preservation, the Montana State Historic Preservation Office, Tri-County Historical Society (Anaconda), Anacondans to Preserve the Stack, and Butte the Uptown Association. Moreover, the plan "grew out of the perceived need to develop a comprehensive and coordinated plan of action to protect and interpret the historic features of Butte and Anaconda. Many of these nationally significant mining and smelting sites are threatened by demolition, vandalism, natural deterioration, and proposed reclamation. . . . This study [is] a first step in combating these problems while integrating historic preservation and environmental reclamation."[104]

This plan stemmed partially from the many requests for support of preservation and community projects received by the Anaconda Mineral Company and from its own desire for a coordinated approach to site management. The master plan provided historic background to the region, several interpretive themes for future heritage exhibits, anticipated community contributions and tourism benefits, and a phased approach to implementation. Although the National Park Service ultimately downplayed the idea of a National Historical Park in Butte-Anaconda, citing a new copper industry–themed national park in Michigan's mining district, regional planners in Montana hoped to create at a minimum a partnership-based National Heritage Area.[105]

While the regional park plan was being worked on, the EPA was in the midst of evaluating the region under CERCLA and determining the best course of action for remediation. The park plan recognized the significant potential impact to historic resources from remediation and that future planning would require "creative thinking and innovative reclamation" to successfully meet both goals of historic preservation and cleanup activities. Despite wide acceptance by stakeholders, however, the plan did not have any enforcement capacity.[106]

CERCLA-mandated remediation projects were well under way by the early 1990s, when the heritage community and the state noted violations of Sections 106 and 110 of the National Historic Preservation Act, demonstrating the difficulties of enforcing the voluntary park plan against the actions of very large organizations and corporations. In most Section 106 or 110 proceedings, the state historic preservation officer monitors and rules on issues of National Register eligibility and on required mitigation of impacts—often on a case-by-case basis. However, the extent of this particular Superfund project, and the extensive potential for impacts, required a more formally planned approach.

In 1992, and again in 1994, the Environmental Protection Agency, the Advisory Council on Historic Preservation, the Montana State Historic Preservation Office, the Montana Department of Health and Environment, the counties of Butte/Silver Bow and Anaconda/Deer Lodge, Walkerville, and ARCO entered into programmatic agreements regarding the creation of a regional historic preservation plan (1992) and an implementation strategy for the completed plan (1994). Recognizing the "substantial potential for the loss of historic resources . . . under CERCLA" and the importance of compliance with NHPA, the parties agreed to create a comprehensive preservation plan to consider potential impacts before they occurred, decide which significant structures would require what type of mitigation when the impacts occurred, and recognize the potential for "undiscovered' resources and mitigation procedures.[107]

The second significant planning document was the Regional Historic Preservation Plan (RHPP), developed by the twenty-seven-member RHPP Joint Committee in 1993. The RHPP capped ten years of heritage planning in the district and brought together the work of all the previous surveys and plans and included sections on preservation and remediation issues, interpretive themes and related sites, and revitalization, management, and planning. It also outlined the goals of preservation articulated by CERCLA and included relevant legal requirements of Sections 106 and 110 of the NHPA, and earlier programmatic agreements and studies.[108]

Lastly, in 2005, the secretary of the interior substantially expanded the National Historic Landmark district beyond its 1961 focus on Butte. The district grew from the original thirty-one individual contributing structures listed in the 1972 inventory to over five thousand.[109] After successive expansions in Butte, the new NHL district designation, organized under new interpretive considerations expressed in the National Park Service labor theme study, included substantial new commercial, residential, and industrial heritage sites in Butte and Anaconda, as well as the railroad corridor between the two. This document, with the studies and plans executed in the 1990s, completed the transformation of two distinct cities into one single, large heritage-planning district. Although no single national or state historic park was established to interpret the region as a whole, both the RHPP and NHL district designation provided a broad coordinated understanding and approach to heritage management in the region.

As the planning process progressed, the communities of Butte and Anaconda both began enthusiastically engaging their history and backing heritage projects, while at the same time the national perception of their region was declining. New organizations such as Butte Citizens for Preservation and Restoration (CPR) and the Restoration Alliance worked to support the efforts of the existing, NPS-sponsored Certified Local Government and Main Street preservation and development programs.[110] But despite the growing strength of the local heritage movement, becoming the second largest National Historic Landmark district in the United States was no match for the stigma of being the one of the largest Superfund sites in the nation. From the 1990s to 2009, the national media focused on the environmental devastation. Articles on Butte-Anaconda appeared with titles like, "We Want People Who Have a Problem with Mine Wastes to Think of Butte" (1992), "Pennies from Hell" (1996), "Butte America: Poisoned, Ruined, and Self-Cannibalized" (1997), "Butte Breaks Ground to Mop up a World-Class Mess" (1998), "Richest Hill on Earth and Costliest Mine Superfund Site" (2000), "Montanans Weigh Options on a Toxic Legacy" (2001), and "In a Town Called Opportunity, Distress over a Dump" (2005). These pieces contributed to existing, and previously cited, designations such as "*ugliest spot in Montana*" (1993) and "*one of the world's must-miss vacation spots*" (2009). Occasionally, some glimmer of hope that all had not been destroyed by pollution appeared in articles like "In its Own Decay, Butte Sees a National Treasure" (1997) and "Town Pins Hopes on Superfund Site" (1997).[111]

Smelter Heritage

Despite the pervasive drive of the Superfund cleanup, three significant elements of the region's smelting landscape remained in 2009: the Butte Reduction Works slag walls, the Washoe Stack, and foundations and waste piles of the Anaconda Old Works. All three of these sites were slated for demolition at some point in the Superfund planning process, but regional heritage advocates organized to save them. The rest of the Washoe smelter site, however, was not so lucky and was demolished just as the Superfund process and more organized heritage planning began in Montana.

The Butte Reduction Works operated from 1885–1910 at the base of Butte hill along Silver Bow Creek, south of town. Early in its operations, like other smelters in the district, the Butte Reduction works realized that existing concentrating technology left considerable copper in the tailings, and, with growing concern about potential lawsuits from downstream land owners, it stockpiled its tailings. Unlike the earthen berms the Anaconda Company used to contain some wastes, the Butte Reduction Works built waste slag retaining walls to impound its tailings and keep them from washing down Silver Bow Creek during spring floods.[112] Although the smelter itself was largely dismantled by WWII, much of the tailings and slag retaining walls remained at the start of the Superfund cleanup (see figure 4.6).

The EPA assigned the two parts of the site Operable Unit *Lower Area One* and Operable Unit *Reduction Works Tailings*, and in 1993 began removing 1.2 million cubic yards of metals- and arsenic-contaminated tailings and other hazardous wastes. The materials were disposed of in a nearby storage area, where they were capped with clean soils and covered with an impermeable membrane to prevent water from leaching acids and heavy metals. The second phase of remediation involved restoring the natural flow of Silver Bow Creek, creating a buffer wetlands area, installing a groundwater treatment facility, and constructing an interpretive trail through the site.[113] According to a federal official, however, the initial plan also called for the removal of the slag walls that ran throughout the site. This effort would have lowered the cost and complexity of the remediation project and contributed to a more "natural" looking area—both high priorities for the EPA.[114]

However, the 1985 park plan called the slag walls at the site "the best preserved remnant of [the Butte smelting component of] the local mining history," and cited its very high integrity and interpretive potential with minimal safety concerns.[115] It also called for an interpreted, self-guided tour through the site.[116] The 1993 Regional Historic Preservation Plan called the slag walls the primary historic feature of the site and asserted that remediation efforts should avoid damaging or removing them as the cleanup progressed, suggesting that this site be used to interpret the early technology of smelting in Butte.[117] In an article from 2000, Fredrick Quivik, who was

also a primary author of the 1985 park plan, referred to the Butte Reduction Works slag canyon as an often-ignored industrial feature that was not simply a worthless waste product, but rather an ingenious way to conserve its other wastes (tailings) and to prevent property damage and potential litigation.[118]

In 2009, the Lower Area One and Reduction Works Tailings project was more than halfway complete, and the slag walls had been saved as an important cultural landscape feature. An interpretative sign standing near one of the walls described the historic use of the site, the cleanup, and the future public use planned for the site.[119] The slag walls in Butte were preserved because of the heritage planning efforts of the county, early recognition and declaration of their significance, and a well-thought-out approach to managing heritage amid cleanup efforts. The two Anaconda smelter sites, however, were saved largely by creative thinking in the wake of significant cleanup momentum.

ACM had largely demolished and removed the structures of the nineteenth-century upper and lower smelter sites in the early twentieth

Figure 4.6. (*Left and right*) Slag walls of the Butte reduction works, 2008, showing former enclosed tailings ponds and close up of unique slag wall construction technique. By the author.

century. Although some historic building foundations and toppled smoke stacks existed on the site in the early 1990s, large extant waste slag and sinter piles from its smelting period generally had long encouraged the residents of Anaconda to use the site as a dump. The EPA estimated the volume of the historic waste piles at over 1.5 million cubic yards of flue dusts, slag, and soils contaminated with arsenic, cadmium, copper, lead, and zinc.[120] Although the Engineering Evaluation/Cost Analysis work plan outlined four pages of actions required to satisfy the NHPA, the General Work Plan called for the stabilization of debris piles to prevent further contamination of Warm Springs Creek during high water events. This effort included consolidating the various piles into a single waste repository, then capping and covering to prevent water leaching.[121]

The 1985 park plan described both the Upper and Lower Works sites as largely foundations strewn with brick, stone, and abandoned waste. Although of high historical importance, the plan noted the multiple locations' low structural integrity and moderate accessibility but still called for a relatively large interpretive center and an interpreted trail along the abandoned rail bed. [122] Although the interpretive center was not built, by 2009 a paved trail wound through the Upper Works site along a former rail bed with signage identifying significant features of the site and others describing the processes and people formerly tied to the site. The rest of the site, with its intertwining ruins and slag and sinter piles, was capped with eighteen to twenty inches of new soil, revegetated, and protected with a new drainage system to prevent leaching into the creek and to provide a stable cover over the waste deposits; it was then converted into a Jack Nicklaus–designed golf course.

During the golf course planning and design phases, Nicklaus held steadfast to some heritage ideals, keeping many aspects of the site's ruins as part of the course. As the golf course stands, the foundations of the Upper Work's smelter building entice golfers along the number three and number four fairways, while golfers on the number seven tee hit off of the top of a slag mound. Further, there are no white sand traps on the course. Nicklaus filled all the traps with granulated slag, now called "sland," from the Washoe Smelter waste dump, claiming it was superior in many ways to traditional trap sand and less expensive to obtain. The sland was free and virtually inexhaustible, given the size of the deposit resting less than one mile away on the other side of Montana Route 1.[123]

Precisely who deserves credit for the idea of building a golf course among smelter ruins as a boost to the local economy and a first step toward rebranding Anaconda an "outdoors" destination is unclear. Several golf course origination stories exist—ranging from a retired smelter foreman's joke to a regional developer's "eureka" moment while playing golf in Idaho. However it started, many claimed that the course was an important defining moment for the community. Just a few years earlier, the EPA had permanently relocated the residents of the neighboring town of Mill Creek because of significant arsenic contamination in homes and local soils, an event that was likely the low point in locals' perception of their landscape. The golf course was designed not only to serve as a control barrier between the public and the hazardous materials underneath, but also to bring 120 full-time jobs and almost nine hundred thousand dollars in income and spending to the community, not including money spent by golfers[124] (see figure 4.7).

Although some in the preservation community bemoaned the loss of the historic character of the old works—and others still await the economic windfall expected from the course—the EPA and ARCO believe it has been a large success and that the site is ready to be removed from the National Priorities List (NPL).

Adjacent to the slag dump in Anaconda, the ARCO subsidiary, Anaconda Minerals Company, began plans for the Washoe smelter demolition less than one year after its 1980 shutdown. In 1982, the company hired Cleveland Wrecking and began actively removing sections of the site. Because of the extensive contamination of much of the debris and waste piles—estimated at 185 million cubic yards of tailings, 27 million cubic yards of slag, and 250,000 cubic yards of flue dust[125]—both Cleveland Wrecking and AMC, ironically, were subject to the Resource Conservation and Recovery Act (RCRA) regulations governing the handling and transportation of hazardous

Figure 4.7. Old Works Golf Course showing Washoe smelter stack in the background and "sland" traps in the foreground. The entire ARCO-funded course was designed to provide a long-term cover for the hazardous wastes "stored" below and some income into the local economy, and to retain some heritage features, such as building foundations and slag piles. By the author.

wastes—the very law some asserted that the company had attempted to avoid by shutting down the smelter just two years earlier. That same year, the EPA also began Superfund investigations and, in September 1983, listed the site on the NPL.[126] Ultimately, the EPA-approved demolition plan for the Washoe Smelter was rather extensive and called for removing all structures, leaving only "massive concrete foundations and bulkheads . . . weathered bedrock, contaminated soil, pulverized brick, and some construction debris."[127] Sanctioned demolition work began in 1983 and finished in 1986.[128]

Despite ongoing demolition and plans for the complete removal of structures, the General Work Plan (GWP) for the smelter site completed under CERCLA was written five years after its NPL listing and, with its instructions for overall site remediation management, included a five-page description addressing the need for a cultural resource assessment.[129] Specifically, the GWP noted the need for a survey to identify sites eligible for the National Register prior to demolition work, recognized that the smelter stack was already listed, and explicitly stated that the survey to address historical, archeological, and prehistoric resources must be completed within 180 days of the order to proceed. Further, it called for consideration of identified and eligible resources during the remedial investigation and feasibility study (RI/FS) and in designing remediation alternatives for the site. Unfortunately for the smelter's heritage, the GWP was finished in 1988, six years after demolition began, and its authors simply assumed that the appropriate Section 106 and 110 actions it outlined had taken place.[130] The RI/FS was finished in 1991 and largely did not address heritage because most of the site was already gone.

No evidence exists to indicate that ARCO or the EPA followed through with the cultural resource survey as they should have under Section 106. In any case, little was ultimately left at the smelter site by 1988 when the General Work Plan identified the need for a cultural resource assessment. Under Section 106, at minimum some sort of documentation *should* have occurred, but no record of this survey or any other heritage study existed in the Superfund repositories in Anaconda, Butte, Helena, or the EPA regional office in Denver. By 1988, what was once claimed to be the largest and most advanced copper smelter in the world had been completely reduced to a single, albeit exceptionally tall, freestanding masonry stack which would became the smallest state park in Montana.

Following the dramatic demolition of the Great Falls stack in 1982, and the announcement of plans to demolish the Washoe smelter, several community members formed the "Anacondans to Preserve the Stack" committee (so named to distinguish itself from the Great Falls "Save the Stack" group) to pressure ARCO and state officials to leave the Washoe stack in place. According to a leader of the preservation movement, the group recognized that to succeed it needed both local and state government support before approaching ARCO. Fortunately, committee members were only about an hour's drive from the state capital and could lobby legislators and attend meetings on very short notice. Although the group's founders were related to the bricklayers who constructed the stack, and the group's initial intentions were to create a monument to the masons, many in the community and the local press enthusiastically backed the plan as a memorial to all smelter workers. According to another preservation leader, the organizers met with little resistance from politicians in Helena, who were eager for any positive development in the area, and the state agreed to create a park. ARCO, for its part, was also enthusiastic about stack preservation, largely because it expected much greater resistance to the smelter demolition and, the community, maybe a little shortsighted, had only asked to save one structure, which otherwise would have also been very expensive for ARCO to remove.[131]

The fourteen-acre Anaconda Stack State Park sits in two parcels. One is simply the in situ stack and a small area around it on smelter hill. Although there may eventually be limited public access or guided tours to the stack itself, in 2008 the hill parcel was restricted because of existing soil contamination and the potential for falling bricks and debris. The second parcel is approximately two miles away and is called the "viewing area." It includes interpretive signs describing the operations of the stack and smelter, data comparing the stack's size to other similarly tall structures, like the Washington Monument and Eiffel Tower, and a full-width, three-dimensional brick model of the stack base that exactly replicates the stack's seventy-five-foot external and sixty-foot internal diameter, while only rising four feet from the ground[132] (see figure 4.8).

The 1985 park plan addressed the initial public outcry over the planned stack demolition, noted that it had been converted into a state park, and largely assumed that any hope of saving any other smelter structures

had been already lost.[133] The plan, however, recognized the significant interpretive potential of the stack and state park viewing area and called for an expanded visitor center and outdoors museum that would include salvaged equipment from the site. Although the state installed interpretive panels describing smelter operations, replicated the stack's cross-section, and placed some equipment and ladles around the park, the formal museum and visitor's center was never constructed.

The Anaconda stack and the Butte gallows frames provide single, simple representational brands—or visual icons—for the heritage of the region. Although most locals embraced its preservation, a few dissenting opinions opposed saving the stack. Some cultural anthropologists might argue that the stack, when operating, symbolized a nonverbal hegemony over the workforce, constantly reminding them of the company's omnipresence

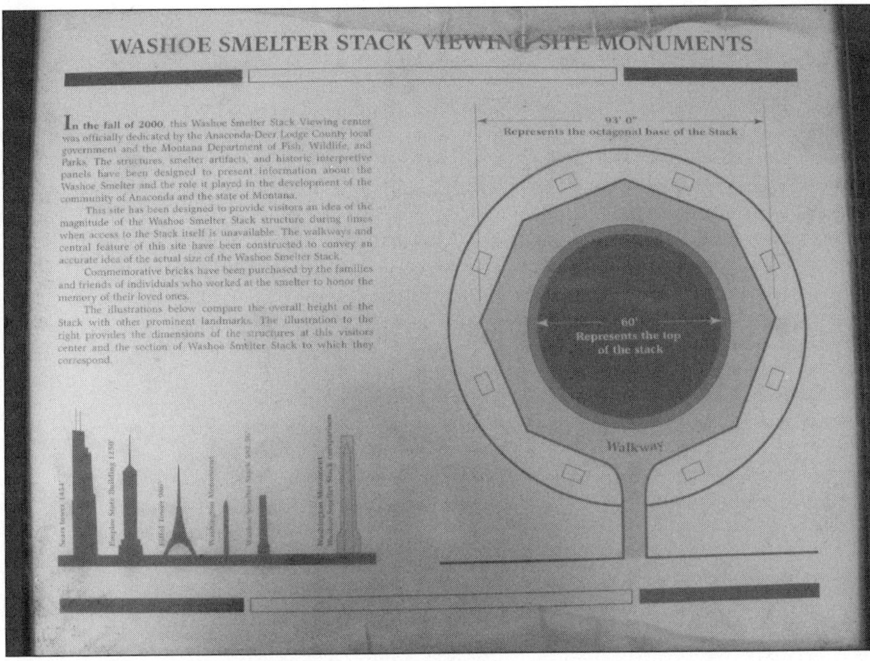

Figure 4.8. (*Left and right*) Interpretive panels at the Anaconda Stack State Park. This smallest state park in Montana sits in plain sight of the world's largest masonry stack, and its interpretive displays focus on smelter operations in general and the stack specifically. By the author.

and ultimate control over their lives. Some economists might claim that the preservation of the gallows frames was tying the city to old economic models and mindsets and preventing it from freely looking forward and planning for the future.[134] In some ways, it may be hard to argue with these interpretations, but with the company gone and heritage tourism a viable economic tool, borrowing these former Anaconda Copper Mining symbols may have greater positive value. In many ways, the stack and gallows frames now stand as reminders of what once gave the place meaning—and of what the region pins its hopes on for the future. Unlike the Berkeley Pit, which serves more as a monument to pollution and to the negative aspects of mining on the environment, the stack and gallows frames, although responsible for the retrieval and distribution of significant pollutants for more than a century, have become icons for positive revitalization.

Almost every person I talked to in Anaconda knows the stack's height, the fact that it is still the tallest masonry structure in the world, and its association with, at one time, one of the greatest copper districts in the world. The same goes for the Butte gallows frames. Although the Butte gallows frames brought toxic materials to the surface and created acid runoff and mine drainage, their preservation is regarded by locals as a monument to the people who worked there—exclusive of the company. The other sites in the Butte-Anaconda corridor try to balance their story between the toxic and heroic, seeking to turn the imagery of human contamination into human triumph over both the land itself and the mess made by the copper industry. Like Ducktown and the Keweenaw, Butte-Anaconda hopes to use historic industrial sites to revitalize its economy and draw as many tourists to the region as possible.

If a single theme runs through the environmental and heritage development of the Butte-Anaconda area it is *largesse*. Historians and locals declare that Butte had the "richest" hill and the most production, and that ACM was the biggest nonferrous minerals company in the world for a time. Anaconda, both the town and the company, had the largest copper smelter in the world, which housed the largest copper reverberatory and blast furnaces, and the tallest masonry structure. The district also had the largest Superfund site in terms of physical size and, costs, and one of the largest National Historic Landmark districts. It has the largest grassroots heritage movement of the three studied in this work and one of the largest heritage planning efforts for a Superfund site.

Although many of these efforts were large in size, the other key feature of the heritage process in Montana was the organization of a strong bottom-up planning effort. Many of the planners were not locals at first, but after arriving, they identified with the landscape and organized and planned for the eventual Section 106 and 110 conflicts with EPA and ARCO with little national-level support except for expanding NHL district boundaries. Butte also demonstrates the value of place and the favorable perceptions of new locals who value the landscape as they first see it. The miners and laborers of the region during the 1950s–1980s witnessed the decline of their district, the loss of neighborhoods, and considerable demoralizing change. New people who moved into the region in the late 1970s and 1980s, outsiders at first, did not necessarily see the loss of the previous decades, but rather saw the considerable wealth of cultural resources remaining, and organized, along with some long-time locals, to protect those enduring resources. Although their efforts produced planning documents and led to important preservation successes, not all resources could be saved—then the newcomers experienced a sense of defeat over the cultural losses, especially as the cleanup gained momentum and brought major changes to the landscape. This population is as likely to talk about its losses as its successes. Yet people new to region in the 2000s, aware of, but not witness to, the losses of the 1950s, 1960s, 1970s, 1980s, and 1990s, but nonetheless interested in heritage, are able to enjoy the results of planning efforts, preservation projects, and heritage integration into the city landscape carried out by local preservationists in the 1980s and 1990s. It is easy to be significantly impressed with the level of what was saved, not with what was lost, which is exactly the sentiment civic leaders want visitors to have.

Similar to the Butte District, the Keweenaw Peninsula saw considerable declines in its mining, which led to efforts to commemorate the past. Unlike Montana, however, where heritage planning was largely organized from the grassroots and dominated by local planning efforts, Michigan heritage planning was driven by a combination of National Park Service, top-down planning, and the efforts of long-standing tourism and heritage groups, all of which melded into a quasistructured historical park.

Chapter 5

Quincy Smelter

> **WHEREAS,** the Quincy Smelting works is the last remaining historic smelter left anywhere in the world and reflects smelting technologies of the late nineteenth Century . . .
> **WHEREAS,** the smelter site continues to severely deteriorate with open windows, roofs, and walls . . .
> **WHEREAS,** the [Keweenaw National Historical Park] has not been successful in securing funds to stabilize buildings nor to restore any of the Smelter buildings for historic interpretive purposes;
> **NOW THEREFORE BE IT RESOLVED**, that the Cities of Houghton and Hancock City Councils . . . do hereby support the restoration of the Quincy Smelting Works and request the Keweenaw National Historic Park make the restoration of the Quincy Smelting Works its highest priority, funded by the Federal FY 2010 budget and
> **BE IT FURTHER RESOLVED**, that if this effort to restore the Quincy Smelting Works is not funded by the Fiscal Year 2010 Federal Budget, the facility be recommended for demolition in the interest of public safety and blight removal, and allow redevelopment opportunities to be undertaken by Franklin Township.
>
> —Houghton Hancock, Joint City Councils Resolution (2007)

The first historic mining-related landscape feature a visitor to the Michigan Copper Country sees and recognizes as industrial is likely a head frame. Most visitors pass dark sand beaches unaware that they're looking at stamp sands and drive through Michigan Technological University not knowing that it is the twenty-first-century incarnation of the Michigan School of Mines, founded in 1885. No—the first physical element that most visitors likely notice is the regionally iconic Quincy shaft-rockhouse that is part head frame and part sorting building. It sits on the apex of Quincy Hill, just up from the Portage Lake waterfront. Like the Butte gallows frames, the clean, reclad shaft-rockhouse is a ubiquitous postmining symbol that appears in many images and forms throughout the region.

But much more like the trip from Ducktown to Copperhill than the trip from Butte to Anaconda, as visitors proceed north along US Highway 41 and approach downtown Houghton, they may glimpse a largely disheveled conglomeration of historic buildings in various states of disrepair on the waterfront across the narrow lake. Although not as large or imposing as the Copperhill site, this collection of structures belongs to the smallest of the big three twentieth-century smelters on the Keweenaw Peninsula and is nearly as intact in 2008 as it was when it closed in 1971, after 73 years of service. Most people in the region desire some sort of positive development of such a valuable and visible piece of waterfront property. Although some would like to see the buildings preserved and interpreted as an important historic site with mixed-use public space, others, citing blight and decay, call for their complete removal to allow for new residential or commercial opportunities. This dichotomous sentiment is echoed in the joint city council resolution quoted above.

Copper smelters and smelting landscapes in general receive only modest heritage attention often because smelters, which create some of the largest environmental problems in mining districts, tend to get demolished and remediated quickly after their functional use is gone. Further, smelters often exist in the historical shadow of their associated mining landscapes, which tend to get much more heritage attention. Although many mining museums talk about smelting, there are virtually no copper smelting museums and, compared to mining heritage sites, almost no preserved and interpreted American copper smelter sites or partial sites exist outside of the fading thirty-acre parcel behind the Ducktown Museum, the Anaconda Stack, and, to a lesser degree, the Old Works Golf Course. Even Swansea, Wales, home to the single greatest concentration of copper smelters ever, pays little homage to—and has few standing remains left of—the historic industry of the town.

The Michigan copper district, however, has an opportunity to change the perception of smelting and its technological, social, and ecological effects on the metallurgical production system. The Quincy is the only near-complete *historic* copper smelter site in the world with in situ reverberatory furnaces and related equipment, offices, support buildings, and *still-occupied* homes of smelter workers (see figure 5.1). The site sits within the Quincy Mine National Historic Landmark district, is a former Historic American Engineering Record (HAER) documentation project, and is a significant contributing site to the Keweenaw National Historical Park (KNHP) (see figure 5.2). Unfortunately, the site is also part of the Torch Lake Superfund project and has variously been considered for full preservation and full demolition. The smelter's current owner, Franklin Township, does not have the resources to invest in the preservation of the site, which has largely sat unattended and unmaintained since its last heat in 1971, slowly succumbing to the ravages of vandals and harsh winters.

Although the site is a contributing location to the KNHP, the National Park Service Washington Headquarters requires careful analysis and consideration before allowing its individual parks to own or get too involved with sites saddled with contamination or other environmental liabilities. Further, in the 1992 enabling legislation that created the KNHP, congress specifically excluded the NPS from acquiring properties that have significant Superfund contaminants. Complicating the preservation of the smelter site, EPA, in its zeal to identify at least one potentially responsible party (PRP) to pay for part of its Superfund cleanup in the area, named the NPS a PRP because the park had assisted the township with preservation and use planning. This move ultimately forced KHNP to step back from assisting the township until the Superfund liabilities were removed.

Shutdown

By the 1930s, both the Tennessee Copper Company and Anaconda Copper Mining Company completely dominated production in their respective districts, controlling almost all mining, milling, and smelting operations. Both companies continued their singular hold over their regions until large, multinational petroleum companies bought them in the 1960s and 70s, and continued to produce copper or related chemicals into the 1980s. Unlike both Ducktown and Butte, however, Michigan copper mining did not coalesce into a single corporate entity. Instead, three companies of the three hundred or so native copper mining companies chartered in Michigan survived until

Figure 5.1. Quincy smelter site, 2008. By the author.

operations in the district ceased. One of the three, however, the Calumet and Hecla, out-lived and out-produced the other two—the Copper Range Consolidated and Quincy Mining Company.[1] Also, although Tennessee and Montana copper mines both worked the same deposits for a century, Michigan's copper mines worked multiple ore types that stretched over one hundred miles in three different Keweenaw Peninsula counties, leading to multiple starts, shutdowns, and the abandonment of various regional mining districts, often leaving ghost towns scattered throughout a still-active mining region.

Although Michigan copper mining production peaked in the late 1910s, its prominence on the world stage began declining in the 1880s, as Butte copper production matched and then surpassed Keweenaw output. From 1847–1880, Michigan copper mines produced nearly 75% of U.S. domestic copper, reaching a high of 82% in 1880, just as Montana silver miners reached the rich Butte copper lodes.[2] Between 1880 and 1884, annual cop-

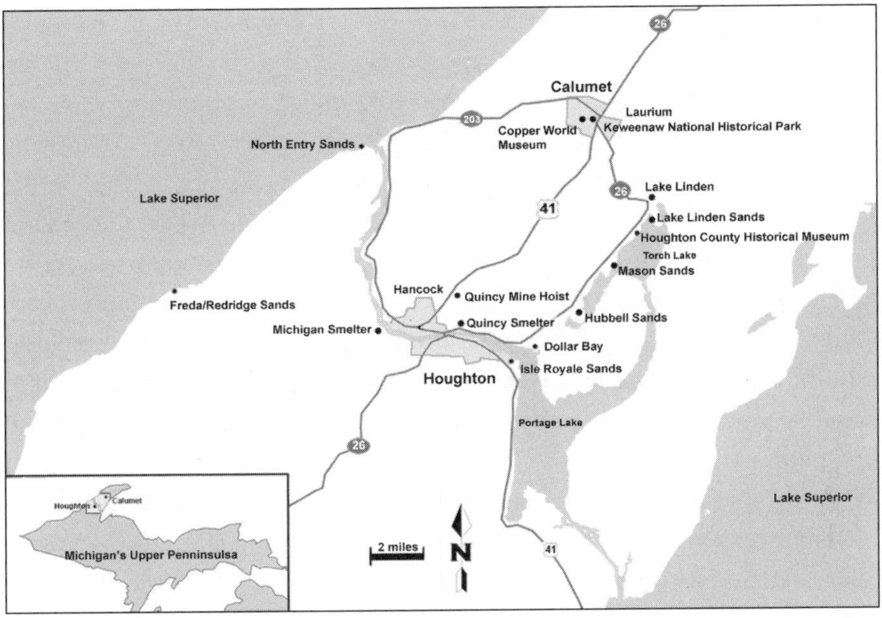

Figure 5.2. Map of the Keweenaw showing selected heritage and remediation sites. By the author.

per production outside of Michigan rose from 10 million pounds per year to 76 million pounds, causing a steep price drop from \$0.20 per pound to \$0.14 over the same period, and a further drop to \$0.11 per pound in 1885. In an attempt to arrest the growth of Montana mines and ultimately stabilize prices, Calumet and Hecla dumped excessive amounts of cheap copper into the market between 1886 and 1887 and forced the price down to \$0.10 per pound—lower, it hoped, than the production costs of many newer western mines. Despite underground fires at C&H mines in 1887, which reduced its production and forced the company to halt its plan to control the market, several small and marginal Montana copper producers went out of business. The plan, however, ultimately failed to affect the Anaconda Copper Mining Company, whose 1887 output dominated Butte copper production and led Montana past Michigan to become the premier copper district in the United States and the world. By 1890, Michigan mines were producing 39% of the U.S. total and by 1900 only 24%. [3]

Keweenaw copper production, however, continued to grow despite mounting complications. As with all mining districts, the remaining Michigan ores got leaner as the richer veins were mined out. Further, Michigan copper mines became some of the deepest mines in the world, following copper veins to distances of over 9,000 feet from the surface by 1920, while the deepest Butte mines were operating at 3,800 feet, and Arizona mines did not have to venture more than 2,915 feet down.[4] Despite the increasing costs associated with deeper mining of leaner ores, Michigan companies were able to take advantage of technologies such as compressed air, electricity, and the one-man rock drill to remain cost competitive and—with simpler processing and smelting—to produce copper cheaper than their western competitors through the early 1910s and continue to pay dividends through WWI. The former production cost advantage of established Michigan mines, already declined to a 10% cost disadvantage in 1916, further declined to a 30% cost disadvantage by 1920 when wartime-mandated price controls ended.[5]

Despite the persistence of profits and the adoption of new technologies, the overall productivity of Michigan mines began declining even before the war. The largest and costliest strike in the district began in 1913, partly over cost-cutting measures such as the one-man rock drill and reductions to workers' wages and hours. Although the mining companies prevailed, workforce problems persisted and even worsened as miners left the area to find better jobs in the west, new immigrants avoided the region because of limited work during the strike, and new, well-paid manufacturing jobs, especially in Detroit, led to a general exodus of workers. Compounding these problems, WWI drew young men into the armed forces, contributing to a cumulative labor force drop from 15,000 mining workers in 1915 to 8,000 in 1920.[6]

Although the postwar economic recession affected the copper industry nation wide, the prewar conditions that were affecting Michigan production worsened in the 1920s. In 1921, many Michigan copper mines temporarily shut down to allow stock piles to be sold off and, in the process, posted their lowest production levels since 1889; also, for the first time since 1847, they did not pay a single dividend.[7] Despite declines and difficult economic conditions, mining continued on the Keweenaw and would continue in some instances for almost fifty more years, but the dominance of the region had clearly faded as the remaining mines introduced new practices and diversified while they struggled to extract the last bits of profit from the local rock.

As discussed in Chapter 2, C&H looked to alternative sources of income and savings and began "mining" its waste tailings in 1915, recovering copper through leaching and flotation processes at half the cost of underground work.[8] In 1919, Quincy began installing the largest steam hoist constructed, capable of winding 12,000 feet of cable and "quickly" moving men, rock, and materials to and from the deepest parts of its 9,000-plus-foot-deep mine. Following depression-era shutdowns and WWII-era production increases, it also began recovering lost copper from its tailings and operated its reclamation mill through 1967 and its smelter through 1971. Although Quincy ended underground mining the day its last government contract expired in 1945, Copper Range and C&H further consolidated existing mines and diversified their corporate holdings to remain in metals production.[9] Copper Range closed its mills and smelters shortly after WWII, and opened the copper sulfide–bearing White Pine Mine in Ontonagon County after the United States entered the Korean War. Although the company maintained a native copper mine on the Keweenaw through 1967 to supply copper to its subsidiary companies, it mined, milled, and smelted White Pine sulfide ores until 1997. C&H, which had purchased and developed a number of profitable subsidiary companies by the mid-1960s, became attractive to corporate takeover. In 1967, Universal Oil Products (UOP) absorbed C&H and, in the midst of significantly declining mining returns and a growing labor dispute, closed the last C&H mine in 1969, ending the era of native copper production in Michigan.[10]

Note on Smelting Sites

Shortly after it closed its Michigan Smelter in 1950, Copper Range demolished the site and recovered any saleable materials. It then contracted its remaining Keweenaw smelting needs to Quincy, leaving behind only the foundations and abutments of its once-important smelting works.[11] Quincy continued to operate its smelter through the end of its tailings reclamation and the closure of Copper Range's last native copper mine (the Champion in Painesdale), both in 1967, and processed scrap copper at the site until 1971(see figure 5.3). When environmental concerns mounted after the state of Michigan began insisting on the installation of pollution control equipment, Quincy simply stopped using their smelter, abandoned the site without scheduling any demolition or environmental reclamation, and transferred ownership to the Quincy Development Corporation, an entity formed to manage or sell the mining company's lands. The smelter site, one of the most complete of any in a heritage context, was documented by the Historic American Engineering Record in 1979, and ultimately donated to Franklin Township in 1999, in lieu of back taxes, for development into a museum in partnership with the National Park Service.

The C&H smelter closed in 1969 along with the cessation of mining and sat vacant until 1982, when Michigan Tech Ventures, a private corporation created by Michigan Technological University (MTU) to generate local business opportunities using and promoting MTU-developed technologies, formed Peninsula Copper Industries (PCI). PCI, with technological support from MTU, adapted a C&H process to recover copper from scrap materials,

Figure 5.3. Quincy smelter, 1926. Reproduced with permission of Michigan Tech Archives and Copper Country Historical Collections.

such as circuit boards and electrical wires, to create copper-oxide chemicals for commercial purposes. PCI moved its operations into the 1912 C&H electrolytic plant at the smelter and used the blacksmith shop and former cupola buildings for storage. Most other buildings on site had been or were ultimately demolished, leaving a patchwork of foundations and disused transportation corridors in 2009.[12]

Of the other smelter sites that once processed Michigan native copper in the Keweenaw, several sat partially developed or as undeveloped ruins in 2009. The Dollar Bay site still retained foundations and a few of the Lake Superior Smelter buildings still stood intact, reused by the Houghton County Road Commission in Hancock. The short-lived 1847 Isle Royale and Ohio smelter site on Isle Royale still had foundations of demolished structures and some abandoned equipment, whereas the Lake Superior Native Smelter Works site, which was converted to a foundry in the early twentieth century, had no visible features left except a heavily altered shoreline. None of the other Michigan smelter sites discussed in Chapter 3 retains any physical features.[13]

Early Environmental Concerns

Unlike the noxious gases produced from sulfide smelting in Montana and Tennessee, which caused visible and severe damages to the environment and personal property, Michigan smelting was relatively benign and, as such, did not substantially injure nonmining property or generate significant early concerns among residents and governments about landscape degradation. In fact, forest vegetation and insect populations, both significantly reduced in Montana and Tennessee because of mining operations, were still at times so thick on the Keweenaw—because of long summer growing days and lack of high pollution levels—that residents in some respects looked forward to the winter fallow season and the availability of ice roads on frozen lakes and rivers to ease overland transportation difficulties.[14] Michigan's copper mining operations, like similar operations in other districts, used significant amounts of lumber to shore up mines and heat residences and dumped its stamp sands in local waterways. Although its coal-fired boilers and smelters produced smoke, none of the ultimate effects was greater than any other coal-powered industrial district, and they were largely accepted as the price of mining from the 1840s until the 1960s, when pollution became an important issue across the country and the EPA began examining levels of more invisible toxins associated with mining wastes.[15]

Origins of the Heritage Movement

All three of the broader regions surrounding the copper mining districts studied in this work became tourist destinations because of their rugged mountainous landscapes, perceptions of purity and pristine wildernesses, and rich histories. However, early tourists shunned Butte and Ducktown as specific locations within the larger destination region because their industrial and urban natures starkly contrasted with the pristine wilderness expected of their surroundings. Historical environmental problems only exacerbated this perception. Smoke-generated moonscapes, enormous settling ponds, dirty streets, and the persistent hum of industrial activity well in to the 1980s repelled most nature-seeking visitors. Although Ducktown had had a chronic environmental stigma since the early Supreme Court case in 1906, and Butte had had a significant national environmental stigma since the 1970s—but a brutish image going back to the turn of the century—the Michigan district was never really affected by degradation issues that impacted tourists. The Tennessee Copper Basin saw significant changes to the barren industrial landscape by the 1980s and Butte-Anaconda was investing in historic preservation and heritage interpretation in the 1990s, but they were still both far behind the Keweenaw Peninsula in attracting recreational visitors.

Michigan copper production began its final decline after WWII. But, unlike Montana and Tennessee, its copper industry did not take on new, more destructive forms to stay alive—like the open pits of Butte and the continued production of copper-based organic chemicals and acids on a very large scale in Ducktown. Further, because of its pure copper deposits, Michigan's smelting industry never contaminated the local environment to the degree of the other locations. Just about the same time that the Michigan copper mining industry started seeing significant declines in the twentieth century, without the negative associations of environmental degradation as in other areas, the automobile and growing middle-class incomes afforded a means for people to escape crowded urban areas, and the Copper Country provided a perfect natural attraction.

Geographically, the Michigan copper district was different, too. Both Tennessee and Butte copper districts were located close to scenic destinations and national parks and, in the case of Butte, on a major intersection of interstate highways. This regionalization enabled automobile travelers to avoid industrial areas despite the local economic development push for tourism dollars in economically declining regions. Michigan's copper district happened to be at the end of the road, and its mines were shutting down at a time of increased wealth and a bit of postwar wanderlust brought on by returning GIs in the 1920s and 1950s. Michigan's copper district was located in an area of scenic beauty and because it *was* the scenic destination, as opposed to near the destination, it drew more people directly in the 1920s–1960s than the much more industrially active Montana and Tennessee copper districts. Further, the Keweenaw Peninsula, not unlike Ducktown, was reachable in a day's drive by millions of people from Illinois, Wisconsin, Minnesota, and lower Michigan.

By the late 1930s, tourism was Michigan's second leading industry after automobile manufacturing; and, with personal incomes increasing, the Upper Peninsula became an important draw for its natural beauty and low population density. Natural beauty alone, however, was not necessarily a powerful enough draw to create a tourism market in the Upper Peninsula. According to Aaron Shapiro, tourism required careful planning, development, management, and packaging by a variety of people and organizations.[16] Early in the century, several Michigan tourism boards created programs specifically to lure Detroit, Chicago, Milwaukee, Minneapolis, and even Cleveland residents to the state. Regionally, they included the Upper Peninsula Development Bureau founded in 1911, whose Cloverland publications touted the virtues of northern Michigan vacations and whose support of the Hartman-Rushton Act (MI) helped pass the 1929 law creating a funded and coordinated tourism-development organization that became the Michigan Tourist Council.[17] A second, more local organization, the Copper Country Vacation League, made up of local merchants, resort owners, and mining company representatives formed in the 1920s to promote the Keweenaw Peninsula specifically.

Although it held some of the same vacation advantages as other parts of the Upper Peninsula and Northern Wisconsin, the Keweenaw remained cooler in the summer and snowier in the winter, with Lake Superior

winds blowing across the narrow land mass. Further, the Keweenaw offered significant access to the Great Lake with over one hundred miles of shoreline, and the opportunity to drive to the end of highway US 41, whose southern terminus sat at Biscayne Bay in Miami, Florida. Enhancing the scenic draw, a fair number of early heritage tours sprang up utilizing the area's abandoned mine sites, ghost towns, and historic forts. To get to Keweenaw County, where many of these were located, drivers had to pass directly through what was left of the declining later Michigan copper district centered in Houghton County, and, during the inter- and postwar periods, several local entrepreneurs created mining-related attractions to lure tourists.[18]

Figure 5.4. Tourism on Keweenaw Peninsula. The person on the left worked at both the Arcadia Tourist Mine and Copperama before coming to the Quincy Mine Hoist. The sign on the right touted the salubrious nature of the region from the 1950s. Both images reproduced with permission of Michigan Tech Archives and Copper Country Historical Collections.

Whereas Montana and Tennessee copper districts used heritage to create a tourism market, Michigan's early tourism market created the first heritage sites in its mining region (see figure 5.4).

The 1926 *Cloverland Tourist Guide* advertised Houghton, Michigan as "the most cosmopolitan community in the Copper Country . . . and the gateway and very heart of copperdom." For tourists, the guide listed the city's good boating, bathing, and fishing, no hay fever, and interesting historic sites nearby, including "Fort Wilkins, the Douglas Houghton Memorial, and famous copper mines, mills, and smelters."[19] Nineteen-thirties Works Progress Administration projects supported the tourism infrastructure, including the construction of lodges, golf courses, and scenic drives. Although these were the very beginnings of heritage tourism, the post-WWII period saw the creation of specific mining-related and heritage attractions on the landscape.

One of the first postwar tourism projects came out of a failed attempt to restart mining in the abandoned Arcadian mine across the lake from Houghton during WWII. Although it ended with the death of its founder in 1943, this venture left behind an unusual horizontal mine shaft one mile off the main highway to Keweenaw County. In 1951, the Arcadian's new owners developed it into the first underground "tourist mine" in Michigan. For over thirty-four years, the mine guided visitors 1,500 feet into the hillside and even garnered recognition from *Life Magazine* in 1962.[20] Copperama, another site that opened some time in the 1950s or 1960s, sat right on the main highway one mile south of Houghton. While documentation on this short-lived attraction is sparse, an existing undated flyer promises a "fascinating, highly educational underground tour at one of Michigan's oldest copper processing sites" located at the Isle Royale stamp mill that had closed in 1948.[21]

One of the longest lived of the mining heritage sites and museums began in 1958, following an American Institute of Mining Engineers (AIME) tour of the Quincy Mine Hoist. The members of the group, captivated by the size of the hoist and recognizing its significance to mining engineering, formed the Quincy Mine Hoist Association with a governing board that included representatives of the local AIME chapter, the Nordberg Manufacturing Company (original makers of the hoist), Michigan Technological University, the Quincy Development Corporation, and the public. The new organization petitioned the Quincy Mining Company for a ninety-nine-year lease on the abandoned structure and was intent on preserving the hoist and associated buildings as an engineering landmark. The Quincy Mining Company provided initial funding through a fifty-thousand-dollar no-interest loan (whose repayment was ultimately forgiven) and a significant eight-to-one matching grant fund.[22] In 1968, the Quincy Mine Hoist Association opened for its first tours. In 1970, the site was listed on the National Register of Historic Places and, in 1979, became the center of a Historic American Engineering Record (HAER) documentation project that included the mining company's smelter, shaft rock house, mills, and other support buildings.[23] The preserved buildings ultimately led to a 1989 National Historic Landmark designation and became one of the key factors in the formation of the Keweenaw National Historical Park in 1992, discussed below.

Although many tourist attractions focused on the region's history, no site had been as historically significant as the Quincy Mine Hoist and none had really had as long an impact on the region's heritage. While recognition that the site could become a tourist destination did not escape the governing board's founders, they understood that the site had greater value than just being an income generator. The preservation of such a significant site contributed directly to the HAER documentation project and National Historic Landmark designation, and, in many ways, reinvigorated heritage planning just as the mining era ended. The region's seemingly untouched historic and archeological sites ultimately lured industrial historians and archeologists, several previously involved in HAER and state industrial documentation projects, to Michigan Technological University. This grouping, which would not necessarily be working together in larger universities, coalesced in the social sciences department and created a unique graduate program in industrial archeology. Further, growing interest in industrial history, the importance of Michigan's copper mining to United States history, and the level of preservation in the region contributed to the formation of the aforementioned Keweenaw National Historical Park in 1992 amid mounting concern about the lasting environmental effects of the district's mines, mills, and smelters.

Federal Heritage – Keweenaw National Historical Park

Although, the Keweenaw National Historical Park falls into the arena of postmining heritage and regional redevelopment, its creation was a direct result of early and ongoing heritage activities generated from the creation of regional tourism in the 1920s and 1930s, the preservation of the Quincy Mine Hoist, and the HAER documentation project of 1979. Congress created the Keweenaw National Historical Park in October 1992, following nearly two decades of studies and plans.

The first attempt to create a regional historical park began in 1972 when Universal Oil Products funded the design of "Coppertown, USA," a large open-air museum intended to rival Greenfield Village in Dearborn, Michigan. The attraction planned to feature mining exhibits, copper products displays, ethnic and cultural centers, festival plazas, arts areas, new hotels, and a seventy-foot-tall miner statue. Although the full concept proved a bit too ambitious for the time, a much more modest indoor museum in the old C&H pattern shop opened with traditional interpretive displays on the region's history and a not-surprising emphasis on C&H.[24]

In the mid 1970s, the Historic American Engineering Record conducted broad surveys of American industrial areas to identify significant historic regions and cultural resources in its realm of interest. In 1977, it followed up a survey of Michigan's Lower Peninsula with a survey of its Upper Peninsula. In addition to identifying iron mining, timber, power, and transportation resources, the survey documented Michigan's Copper Country, identifying a significant number of intact mining resources, including the fairly complete system of worker housing, and the mine, mill, and smelter sites of the Quincy Mining Company. Because of the high integrity of the Quincy sites, which included the mine hoist, HAER scheduled a full site documentation of the Quincy for 1979. That summer a large group of architects and photographers recorded the site's physical features, and a group of historians wrote its technological and social history, providing a copious amount of formal historical research justifying the region's significance and fueling the push toward a national park.[25]

The first plan for a national historical park on the Keweenaw surfaced in 1974 but did not garner the political support it needed and died shortly thereafter. The idea, however, reemerged and gained real momentum following a governor's tourism conference in 1986. At the time, supporters reintroduced the concept of a national park to arrest the growing decay of Calumet, once the center of the region's copper mining industry, and began to pressure Michigan's influential congressmen for backing. This course of action led to a congressional request for consideration, and the NPS researched and published *Options for National Park Service Involvement in the Management of Historic Copper Mining Resources on Michigan's Keweenaw Peninsula* in 1988. Although it neither recommend nor opposed a park, this report demonstrated the feasibility of a park and, if a park were to be established, proposed two units, one at Quincy anchored by the existing mine hoist and one in Calumet centered on the Coppertown Museum.[26] The report also articulated the need for National Historic Landmark (NHL) district nominations for the two areas to demonstrate the high integrity and national significance required for a national park.[27] The *Options* report and subsequent NHL nominations led to a study of alternative designs, which examined a variety of potential park arrangements and ultimately recommended a two-unit park managed under cooperative agreements with the NPS. The federal government under the management of new park would actually own just a few key buildings and relatively little property. They would, however, support affiliated sites both within the park boundaries and related sites outside the boundaries that agreed to operate and run programs to certain standards and promote historic preservation and adaptive reuse of buildings and landscapes.[28]

Congress passed Public Law 102-543, creating the Keweenaw National Historical Park (KNHP) in 1992 to "tell the story of the role of copper in the development of an American industrial society and the effects on the Keweenaw Peninsula of providing the copper" and "identify, study, and preserve significant historical and cultural sites, structures, districts and landscapes . . . for the education, benefit, and inspiration of present and future generations . . . and develop and sustain into the 21st century the park and the community through a blend of private, local, state, and federal management, investment, and ownership."[29] KNHP officially came into existence with the signature of the president, and the NPS in Washington allowed the new unit to proceed with land acquisitions, staff building, and limited financial assistance to partnership sites. Although KNHP had a fair degree of autonomy in deciding how to proceed, one important clause in the law, section 4 (b), restricted the park from acquiring lands that were contaminated with hazardous substances.[30] This clause forced KNHP to explore only those parts of an industrial landscape that had not housed industrial activity or that might have produced hazardous wastes, such as mines, mills, and smelters.

Thus, the Keweenaw National Historical Park owns only five buildings in the park, including former administrative, warehouse, streetcar, and commercial facilities, but does not own any site where industrial processes occurred. It originally partnered with twenty-three existing privately owned and operated historic sites, a reflection of the region's early involvement in heritage tourism, to interpret the region as a whole. Although several of the sites have gone out of business over the years, by 2008 the list of heritage partner sites in the region included the Adventure Mining Company,

A. E. Seamen Mineral Museum, Calumet Theatre, Chassell Heritage Center, Copper Range Historical Museum, Coppertown Mining Museum, Delaware Copper Mine, Finnish American Heritage Center at Finlandia University, Fort Wilkins State Park, Hanka Homstead, Houghton County Historical Museum, Keweenaw County Historical Society, Keweenaw Heritage Center at St. Anne's, Laurium Manor Mansion Tours, Old Victoria, Ontonagon County Historical Museum, Porcupine Mountains Wilderness State Park, Quincy Mine & Hoist, and the U. P. Firefighters Memorial Museum.[31]

Superfund

Concurrent with the formation of KNHP, the Environmental Protection Agency was exploring the very same landscape for hazardous mining wastes from a century of past activity. However, like many remediation projects that ultimately get cleaned up under Superfund, environmental investigations began with reports and health concerns generated by the state. The basis for early state investigations began with a Michigan environmental law designed to curb pollution. Michigan's revised constitution of 1963, Article 4, stated that "the conservation and development of the natural resources of the state hereby declared to be of paramount public concern in the interest of the health, safety, and general welfare of the people. The legislature shall provide for the protection of the air, water, and other natural resources of the state from pollution, impairment, and destruction."[32] Although this declaration mirrored developments nationally and seemed to suggest that human welfare was of equal importance to the future exploitation of natural resources, the first laws actually restricting pollution and implementing Article 4 did not surface until the Michigan Environmental Protection Act of 1970 and the state's clean air and clean water acts, which were written and implemented in the 1970s.

Although Superfund would become the official agent affecting the clean up of Michigan copper mining wastes, state of Michigan departments drove the initial investigations of contamination in the area. In 1973, the Michigan Water Resources Commission investigated a 27,000-gallon cupric ammonium carbonate discharge that had occurred at Lake Linden between 1971 and 1972. Although no "deleterious effects" had been reported from the spill, it raised awareness of other potential mining contamination and prompted state investigations into regional fish abnormalities called locally the "Torch Lake Tumors," and the long-term deposit of xanthates and creosotes from milling operations. Further, in 1985 an International Joint Commission designated the Torch Lake basin a "Great Lakes area of concern" because of potential contamination. Although many of the fish tumors had disappeared by the early 1990s, high levels of mercury and PCBs were found, and fish consumption advisories remained in effect in 2009.[33]

Michigan ultimately ceded control over the abandoned waste remediation to the federal government, despite the fact that the first formal implementation of mining-related pollution controls in the Keweenaw had stemmed from the state and affected the organization and construction of the White Pine mine, mill, and smelter in the 1950s and the final closure of the Quincy Smelter in 1971. The state investigated or supported investigations into copper tailings and slag wastes, which thus contributed to the official listing of Torch Lake on the Superfund National Priorities List in 1986. After that event, the EPA became the primary governmental entity to formally document past environmental conditions, plan for remediation, and attempt to seek a responsible party to pay for the cleanup under CERCLA. The EPA officially took charge but, as in most Superfund cases, studies and remediation plans were developed in consultation with the Michigan Department of Environmental Quality.

Unlike the Butte and Ducktown districts, the pollution problems from the Keweenaw mining operations were relatively benign and not considered a high human health threat. The EPA's classification of the region as an environmental risk related mostly to the deposition of tailings, slag, and airborne dusts, and to ancillary industrial activity such as chemical leaks and discarded storage drums. According to the Torch Lake Second Five Year Report:

> Torch Lake was the site of copper milling and smelting facilities and operations for over 100 years. The lake was a repository for all mining related waste. . . . Copper was extracted through a series of technologies . . . first by crushing or "stamping" the rock . . . then by grinding. . . . The copper and crushed rocks were separated by gravimetric sorting in a liquid medium. The copper was then sent

> to the smelter. The crushed rock particles called "tailings" or "stamp sands" were discarded along with mill processing water, typically by pumping it into the lakes and streams. . . . Dredges were used [starting in 1916] to collect submerged tailings which were screened, re-crushed, and gravity separated. An ammonia leaching process involving cupric ammonium carbonate was used to recover copper. . . . During the 1920s chemical reagents were used to further increase the efficiency of reclamation . . . [and] included lime, pyridine oil, coal tar creosotes, wood creosote, pine oil, and xanthates. After reclamation . . . chemically treated tailings were returned to the lakes.[34]

Over the course of a century, more than 200 million tons of tailings and reclamation wastes were dumped into Torch Lake, filling approximately 20% of its volume. Although national prohibitions against creating obstructions to navigable waterways were established by the Rivers and Harbors Act of 1890, Torch Lake, at the time, was specifically exempted. The lake was further exempted from dumping laws when mining increased during WWII.

During early site investigations conducted under CERCLA, the EPA discovered empty drums at mill and smelter sites along Torch Lake, and over eight hundred drums dumped on the lake bottom. Although those drums determined to have once contained hazardous substances were removed and disposed of at a hazardous waste landfill, many of the innocuous, empty drums on the lake bottom were simply left in place.[35]

Curiously, the EPA did not aggressively pursue potentially responsible parties (PRP) at this Superfund site despite the fact that Universal Oil Products (UOP), Quincy Development Corporation, and the Copper Range Company still existed at the time of the NPL listing. The EPA began exploring cost recovery options in 1986, but, following unsubstantiated claims from UOP that the EPA would make local private landowners fund the cleanup or that UOP would sue local land owners if it were required to pay, the EPA withdrew its pursuit in 1988, claiming it was doing so to assuage local fears about protracted federal lawsuits. A more significant factor in deciding not to pursue PRPs may have been that the total cost of the Torch Lake cleanup was estimated at a mere $15.2 million as opposed to over $1 billion in Butte-

Anaconda and likely several hundred million dollars in Ducktown, and that the cost of pursuing UOP, Quincy Development, and Copper Range would have incurred attorney fees and time costs much higher relative to the cost of the cleanup.[36] Also, the contamination was not a significant human health risk, just an environmental risk, and thereby a lower priority cleanup. In the long run, the EPA may have simply found it less costly to pay for the work from the "super-fund" generated from petroleum and chemical taxes.

Although many of the documented incidents of hazardous substances in the Keweenaw region were relatively minor, the primary concern and target for remediation came from existing tailing and slag waste deposits. The following passage comes from the First EPA Five Year report on Torch Lake:

> The continuous release of tailing- and slag-borne contaminants via wind, surface water runoff, and wave erosion are deemed to represent an unacceptable and actionable source of ecological risk. The most severe ecological impact is the degradation of benthic communities (bottom dwelling organisms) associated with contaminated sediments in Torch Lake and other water bodies at the site. The benthic community is an integral part of the base of a complex food web in lakes. A severely impacted benthic community would impact the entire food web. Toxic effects due to heavy metals (especially copper) appear to be related to sediment pore space dynamics and seem not to have significant water column impacts.[37]

The general plan for remediation was thus to arrest the flow of tailings- and slag-generated toxins into the air and water, and the EPA identified Torch Lake itself and thirteen specific sites on Torch Lake, Portage Lake, and Lake Superior that had large waste deposits. The EPA ultimately broke the sites into three operable units to manage. The plan, according to 1992 Record of Decision (ROD) was to:

> 1. Reduce or minimize potential future risks to human health associated with the inhalation of airborne contaminants from the tailings and/or slag located at the site; 2. Reduce or minimize

> potential future risks to human health associated with direct contact with and/or the ingestion of the tailings and/or the slag located at the site; 3. Reduce or minimize the release of contaminants in tailings to the groundwater through leaching; and 4. Reduce or minimize the release of contaminants in tailings to the surface water and sediment by soil erosion and/or air deposition.[38]

Torch Lake and the thirteen specific locations were divided geographically into the three operable units, shown in table 5.1.[39]

TABLE 5.1. List of Operable Units in Torch Lake Superfund Site

<table>
<tr><td>OPERABLE UNIT 1
10. Mason Sands
11. Hubbell/Tamarack City
12. Lake Linden Sands</td><td rowspan="2">OPERABLE UNIT 3
1. Calumet Lake
2. Boston Pond
3. North Entry
4. Redridge*
5. Freda*</td><td rowspan="2">
6. Michigan Smelter
7. Quincy Smelter
8. Isle Royale Sands
9. Dollar Bay
13. Point Mills</td></tr>
<tr><td>OPERABLE UNIT 2
Torch Lake*</td></tr>
</table>

*No action taken

The ROD called for no action regarding OU2 (Torch Lake) because the lake was recovering on its own, and the general plan of remediation for the majority of other sites was a four-part system of removing debris, grading the tailings and slag deposits, adding a six- to eight-inch soil cap and vegetative layer to arrest blowing and erosion and prevent public exposure, and finally implementing deed restrictions to ensure the remedy remained intact.[40] There were, however, a few exceptions to this treatment besides Torch Lake. Portions of the Isle Royale Sands southeast of Houghton were being redeveloped for a sewage treatment plant, and some of those stamp sands were used in concrete block manufacture. Part of the Point Mills sands were being used for winter road traction, and sands at Freda and Redridge were exposed to heavy Lake Superior wind and wave action and exempt likely because of concerns for the long-term stability of the soil and vegetative cap.

Except for the Quincy site, none of these locations had significant standing historic structures eligible for the National Register, thus Sections 106 and 110 issues were not applied to the EPA's work despite arguments that the stamp sands themselves could be considered a significant historic landscape feature representing major man-made changes to shorelines.[41] The lone exception was the National Historic Landmark–designated Quincy Smelter site, built on stamp sands created by the earlier Pewabic and Franklin mills. The following passage comes from the 2008 Proposed Plan for ROD amendments:

> The Quincy Smelter . . . includes 25 buildings, which historically operated as a copper smelting facility during the nineteenth and twentieth centuries until 1969. The Quincy Smelting Works is a significant component of the Quincy Mining Company National Historic Landmark District established in 1989. This district is also within the boundaries of the Quincy Unit of Keweenaw National Historic Park. Franklin Township owns the Quincy Smelter complex and is interested in developing and restoring the area as part of the national park. Activities conducted under EPA's 4 Superfund Redevelopment Initiative (SRI) have identified Quincy Smelter as having passive recreational use and historical representation (historic site) use.[42]

The smelter soon became a contentious issue for the region. The EPA and Michigan Department of Environmental Quality (MDEQ) claimed that the stamp sands on which the smelter was built were prone to airborne disbursement and that its slag piles contained high quantities of heavy metals; they thus considered demolition as a potential early remedy.[43] Local developers claimed that the deteriorating buildings were an eyesore and that prime waterfront real estate was being wasted; they argued that the buildings should therefore be demolished.[44] The National Park Service argued that the smelter buildings were not only part of a National Historic Landmark district, but they were also affiliated with the Keweenaw National Historical Park and part of the one of the last historic copper smelter complexes in the world and represented a significant component of the local industry and should therefore be preserved.[45] The State Historic Preservation Office, generally in charge of Sections 106 and 110 issues, largely deferred to the NPS.[46] Meanwhile, locals split among preservation, commercial development, and complete demolition.

Quincy Smelter

When selecting Keweenaw sites to be documented by HAER, choosing properties to include in the National Historic Landmark district, and defining the boundaries of the Keweenaw National Historical Park, heritage professionals in each case included the Quincy Smelter site on Portage Lake for its integrity as a historic location and significance to regional and ultimately national copper production. While other significant contributing sites such as Coppertown Museum and the Quincy Mine Hoist were under fairly stable ownership, the smelter had sat largely unused and abandoned since its closing in 1971, suffering annually from the insalubrious affects of heavy winters and chronic vandalism. Unfortunately for the site, the KNHP implementing legislation forbade the new national park from acquiring the site because of potential hazardous wastes and its listing on the NPL. The Quincy Smelter is the only Keweenaw site listed specifically on both the National Historic Landmark District form and the National Priorities List.

The EPA and MDEQ evaluated Keweenaw stamp sand and slag deposits, like those at the smelter location, based on the potential for redevelopment and likely exposure to hazardous substances, primarily antimony, arsenic, copper, chromium, and vanadium. Although most of the deposits had acceptable thresholds for temporary or short-term exposure if they were redeveloped, for example, as parks or recreation areas, the exposure threat became much more problematic if people were going to be spending considerable time near the deposits, if the deposits were to be disturbed, or if children could be at risk of ingestion, which was primarily a concern if the areas were redeveloped into housing. Because of this high risk, the EPA instigated the soil cap and vegetative cover remedy for stamp sand deposits that had the potential for new housing.[47]

The stamp sands at the Quincy Smelter tested high for copper, antimony, and chromium, which contributed to its rating as a chronic hazard for prolonged exposure. Copper was singled out as the "most pronounced contaminant" for children *residing* at the smelter site, but antimony was also rated a high health risk for children.[48] Further, the slag piles also tested high for cadmium and chromium[49] (see figure 5.5). However, unlike the Torch Lake stamp sands, the smelter site from the beginning of the Superfund process had been considered for potential inclusion in the Keweenaw National Historical Park with acceptable exposure risks for visitors and staff.[50] In the 1992 Declaration for the Record of Decision, the EPA stated, "Assuming that the slag pile located in the Quincy Smelter Area (approx. 25 acres) will be developed as part of a National Park, no action will be taken. If the area is not

Figure 5.5. View of Quincy slag pile, 1978. Courtesy of the U.S. National Park Service. Jet Lowe, photographer, "View along trestle for electric trolley leading from reverberatory furnace building to slag dump." HAER MICH, 31-HANC,1-214.

developed in the future, then deed restrictions will be sought to prevent the development of residences in the slag pile area."[51]

In 1992, the same year the Keweenaw Park was formed, the EPA decided that no action was required at the smelter site as long as it was developed as a historic attraction without individual/personal long-term exposure to hazardous wastes. Even if the smelter site was not developed, the EPA decided it would only seek deed restrictions before considering taking the site off the NPL. Unfortunately, the enabling legislation for the park prevented federal ownership of the site, but not technical and financial assistance provided to the site owner. In 1999, the KNHP worked with the site owner, Quincy Development, to transfer the smelter to Franklin Township with the intention that the NPS would provide planning and technical assistance and, at some point in the future, lease or otherwise develop the site if legally permitted. In addition to funding smelter documentation and research projects, and providing technical expertise, restoration cost estimates, and planning concepts to the township over the next five years, KNHP provided guidance to the EPA in the erection of a security fence to protect adjacent trail users from friable and blowing asbestos.[52]

Despite the EPA's 1988 decision not to pursue potentially responsible parties and its expectation, if not support, for the inclusion of the smelter site in the KNHP, EPA and MDEQ interpreted the expenditure of NPS planning funds and its attempt "to impose restrictions on the location of protective fences" as signs of tacit ownership, and the agency's enforcement branch went after KNHP as a new PRP.[53] In a modified form letter received by the KNHP on June 28, 2004, the EPA's Region 5 Emergency Enforcement and Support Section stated that the EPA was planning to upgrade security, prepare health and safety plans, perform comprehensive asbestos sampling, and remove asbestos and any other hazardous substances it found at the Quincy Smelter site. Further, the letter claimed to have received "information that [KNHP] owns or operates the facility or generated or transported hazardous substances that were disposed of at the facility," and encouraged the park "as a potentially responsible party, to agree to reimburse the U.S. EPA for costs incurred to date and to voluntarily perform or finance the response activities which the U.S. EPA has determined or will determine are required at the facility."[54]

Although this effort appeared to have little merit based on the intentions of the NPS, it caused an immediate cessation of action by the local park staff and an order by the NPS regional director to have no further direct involvement with the site beyond technical advice, leaving the care and future of the site to Franklin Township, a governmental unit with little excess revenue to fund the preservation and redevelopment of a twenty-five-structure National Historic Landmark site listed on the National Priorities List.[55]

The site languished under a perplexing cloud, with the EPA both supporting and subverting potential restoration and redevelopment. Although the EPA was ready to take "no action" and delist the site in the 1990s, the state of Michigan DEQ kept insisting that the EPA work to remediate known toxins in the stamp sands and slag, and remove asbestos discovered in the early 2000s. The EPA did not initially pursue full-scale remediation actions because of the potential for a PRP, such as the NPS, to pay for the cleanup, but it did remove hazardous materials stored in drums and ultimately abated asbestos at the site while awaiting resolution of the National Park Service role.[56] The NPS, in the mean time, could not contribute to restoration planning because of the threat of liability, and Franklin Township could not apply for Brownfield, coastal zone management, or other preservation grants because of the NPL designation. Meanwhile, the site further disintegrated with each passing winter, leading to public controversies reflected in the joint city councils resolution cited at the beginning of this chapter.

Like each case studied in this book, public controversy accompanied this Superfund project from the beginning and focused on the pending decision to completely remediate the smelter site or save it. At early public meetings, some residents commended the attempts to preserve the smelter and criticized the efforts to cover other stamp sands, recommending that the "EPA establish a defined protocol for the protection of these remarkable resources whenever they are encountered in any remediation activity."[57] Others clearly admonished the EPA for considering an option not to remediate the smelter site and stamp sands because the potential for any toxicity must be addressed.[58] The EPA responded, that by law it was "required to consider all applicable or relevant and appropriate requirements (ARARs) in selecting and implementing the remedial action. National Historic Preservation Act and Historic Sites, Buildings, and Antiquities Act are some of the ARARs to be considered. . . . U.S. EPA will work with the MDNR (Michigan Department of Natural Resources), local townships, and necessary agencies to protect the historically valuable resources. In addition, based on the development of certain areas as a historical park, the U.S. EPA has excluded those areas from the selected remedy."[59]

The lack of development or preservation activity with the smelter site after more than a decade of discussion encouraged a group of local citizens to form a support group initially called Save Quincy Smelter, but later formalized into the Quincy Smelter Association. Conceived and formed by nonheritage professionals, the group initially called for the development of a shopping area and crafts center in the restored buildings. Although this approach was not enthusiastically received by other stakeholders—especially local city managers whose struggling downtown business districts could use less competition—the controversy served to increase discussion about the site in the community and to force other stakeholders to find more reasonable solutions that could satisfy the desires of the community and historic preservation needs of the site. Although it eventually backed away from the shopping development concept, the Quincy Smelter Association held several fundraisers and supported small projects around the site while serving as a public voice for the community.[60]

In 2006, new inspections demonstrated that stamp sands at the smelter site continued to erode into the lake.[61] Further, the majority of the eleven superfund sites selected for remediation actions had been officially remediated and, with deed restrictions in place, became eligible for delisting from the NPL and available for redevelopment.[62] These actions, largely accomplished by 2005, pushed the EPA toward finalizing the last remediation plans for lingering sites like the smelter and drawing the Torch Lake Superfund Project to a close.

But no formal activity to expand the national park to include the smelter site had occurred, and MDEQ continued to push for remediation projects, while the EPA began exploring its options to finish work at the smelter. With the intent of ultimately amending the ROD regarding the smelter site from "no action because of National Park formation" to some remediation to meet its primary objective of eliminating existing hazards and the requests of MDEQ, the EPA began formalizing final remediation efforts. However, reacting to the joint Houghton-Hancock City Councils resolution calling for demolition of the smelter if no park partnership developed (cited at the start of this chapter), U.S. Senator Carl Levin (Mich.) convened a series of stakeholder conference calls in 2007 with Michigan Department of Environmental Quality, Franklin Township, the Natural Resource Conservation Service of the USDA, Michigan Technological University, and the Keweenaw National Historical Park specifically to engage all parties interested in the site and to find a lasting solution to preserve and remediate the site.[63] Curiously, the Michigan State Historic Preservation Office neglected to join the conversations, and continued to largely rely on NPS for the enforcement of Sections 106 and 110 as it had throughout the whole Superfund process.

Several key decisions came from these meetings. First, it opened a dialogue between the National Park and EPA regarding the issue of Superfund liability, ultimately resulting in a revocation of the EPA's 2004 PRP claims against the NPS.[64] Second, it allowed Franklin Township and key heritage experts at the National Park and MTU a voice in the ultimate solution for the site. Third, it encouraged direct discussions with the EPA and regional stakeholders interested in preservation and, with positive progress, Senator Levin, a long-time supporter of Copper Country preservation projects, identified funding in the national budget specifically earmarked for the smelter site. Fourth, with demonstrated preservation activities focused on the future, Copper Country Preservation, Inc. (CCPI), a newly formed nonprofit organization wholly owned by the KNHP advisory council and organized to support the park's mission and partner sites, decided to support the smelter and local appropriations as its first fund-raising project. In addition to a widespread letter-writing campaign supported by the National Parks Conservation Association, CCPI successfully lobbied the cities of Hancock and Houghton each to pledge five thousand dollars in 2008 for the restoration of the smelter site, an action called for in its 2007 resolution, which had also identified demolition as a potential option.[65]

The EPA, fully aware of the historic nature of the site from the beginning and anxious to finish the project, now decided to remove the exiting slag piles and grade, cap, and vegetate all the exposed stamp sands outside the fence that the EPA erected in 2004 to appease MDEQ (see figure 5.5).

But with strong evidence citing the chemical stability of slag, including the Old Works Golf Course slag traps in Montana, the EPA, following input from NPS and MTU, decided to leave the large slag piles in place with some form of passive deterrence to prevent direct public access and to remediate the rest of the areas outside the fence line, much like other stamp sands in the district.[66] The revised plan was presented to the public at a January 2009 meeting with a follow up presentation of final design alternatives and ROD amendment in late July 2009[67] (see figure 5.6).

Although the smelter site is a long way from restoration, the decision to pursue preservation rather than demolition and new construction seems to be fairly well accepted by the community. In late June 2009, Senator Levin announced an additional one million dollars for the smelter ear-

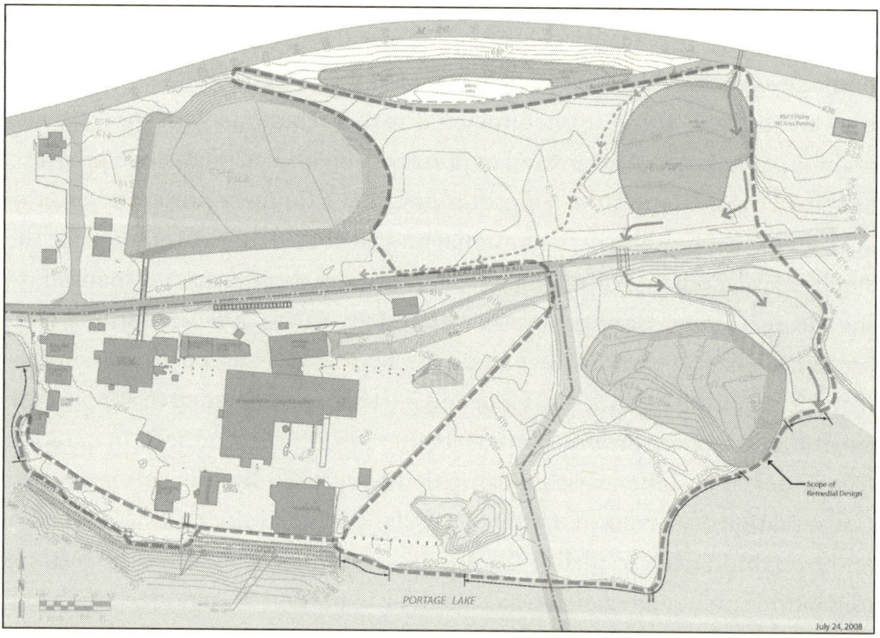

Figure 5.6. Quincy smelter base map for proposed final remediation in 2009. The area inside the dashed lines will be graded, covered with soil, and vegetated to arrest erosion. The slag plies will be left alone, but some form of passive restraint will be erected, such as a short fence. The arrows indicate new water drainage from the hillside. The buildings will not receive any further remediation treatment beyond the asbestos removal in 2008. Courtesy of the EPA.

marked in the 2010 federal budget.[68] This appropriation, however, still had to be approved by the full Senate and survive the Senate/House conference committee and, even if approved, won't be able to be spent for several years.

If the Ducktown district is defined by the landscape and Montana by its largesse, then Michigan can be seen as the most diverse and dispersed. Its mines were spread over three counties and one hundred miles; it had three large mining companies in its eras of high production and decline instead of one. It had the deepest mines and simplest processing. It had the most diverse stakeholders involved in cleanup and heritage conflicts, including the EPA, NPS, MTU, the state of Michigan, many different cities, three counties, and nineteen contributing heritage sites. Its overall heritage planning was a mix of top-down planning coming from the National Park Service and the EPA, with only limited involvement from the state of Michigan, and, of course, bottom-up planning coming from the many heritage sites, the grassroots groups that advocated the national park designation, and the Quincy Smelter Association. Because of the partnership nature of the KNHP and its limited staff size, the park does not impose much control over the partner sites and provides only limited technical and financial assistance, thereby encouraging and necessitating the involvement of grassroots groups. However, NPS does have the resources and bureaucratic infrastructure to consider district-wide planning initiatives, unavailable to and likely unnecessary for individual sites, and the legal department of a federal bureaucracy to respond to letters encouraging it, as a PRP, to pay for environmental restoration. So, by design, the heritage profile in the Keweenaw is a mixed grouping of top-down and bottom-up planning.

The Superfund project had only two direct impacts to the heritage of the region. First, it covered a considerable area of stamp sands that some considered benign and an essential part of the region's character. Second, although it wanted to walk away from the Quincy Smelter in 1992 (and not expend any remediation funding), the EPA was forced to find a compromise solution between a slowly developing historic preservation plan advocated by a United States senator and the MDEQ's desires for full remediation. Although it does not address either concern fully, the solution does respect and respond to the most important components of both sides: protection of the historic structures and slag piles as defined by the National Historic Preservation Act and a cap-and-cover solution over a substantial portion of the stamp sands to remediate questionable health dangers, as defined by CERCLA.

Conclusion

Two key factors combined to drive the dramatic technological change in American copper mining and smelting in the nineteenth century. First, copper demand increased at such a rate in the industrial and modern periods that supply lines often required significant modifications to existing technologies to achieve ever-greater output and efficiency. Second, expansive lean and sulfur-rich lodes were discovered that required complex mechanical, chemical, and thermodynamic processes to economically extricate the copper from their ores. American copper demand was so great, and many of its primary ores so lean, that the industry further required massive corporate organizations and labor forces to ensure profitable scales of operations. Important nineteenth- and twentieth-century copper mines operated in more than twelve states.[1] But such extensive copper recovery from ores, largely between 0.5% and 5% rich, meant that mining, milling, and smelting generated a considerable amount of waste, ultimately leading to significant environmental problems and significant Superfund remediation projects managed by the U.S. Environmental Protection Agency in, among other places, Montana, Tennessee, and Michigan.

By 2008, copper production in the United States was largely centered on very low-grade porphyry deposits in a few mining districts in Arizona, Utah, New Mexico, Nevada, and Montana, and processed at one of only three smelters in Arizona or Utah or outside of the country. Although it no longer had the largest copper output in the world, the U.S. still produced 1.25 million metric tons of copper and processed scrap in 2008, which was a little above the yearly U.S. average for the preceding century.[2] Montana copper production, the smallest of five producing states, came from Montana Resources, Inc. in Butte, and a small portion of the total scrap handled in America came from Peninsula Copper Industries reprocessing of discarded copper-rich components and wire at the former Calumet and Hecla Smelter site in Hubbell, Michigan.

Both the 2008 Montana and Michigan copper operations were relatively insignificant reflections of what the industry once meant to the economies of both regions, but in some small measure, the once-dominant regions still produced copper-rich materials using means and methods developed decades earlier. The economic and technological continuity with the past contributed to the heritage and identities of both regions, especially as they neared the end of controversial Superfund remediation projects. Although Tennessee had no active copper-related industry left after its last chemical plant closed in 2008, it was also nearing the end of active remediation.

The Superfund projects in each district defined the end, perhaps the bitter end, of their once-dominant mining periods. Although two districts created noteworthy by-product processes, all three developed significant mining, milling, and smelting processes on a scale that created jobs, social structures, regional identities, waste streams, and significant wealth, the last one mostly for outside investors. Whereas there tended to be a fairly fluid exchange of mining workers and engineers among active mining districts, later and older generations tended to stay put, especially as production ended. Those later residents had to face the ultimate decline of production and the indignities of not only deepening mineralogical insignificance, but also the implicit culpability of being party to serious environmental degradation without necessarily sharing in much, if any, of the wealth generated by mining.

This study examined the effects of environmental remediation on the heritage and preservation of industrial regions, specifically examining copper smelting in a broader mining landscape and the complex negotiations undertaken and compromises reached by *outside* remediaters and *inside* preservationists, as each argued for what it considered was the best course of action. Chapters 1 and 2 explored the technological and environmental history of copper smelting from its prehistoric origins through the twentieth century, when American copper mines and smelters dominated world production, and the American economy dominated world consumption. Chapters 3 through 5 focused on three of the most significant nineteenth-century copper mining districts in the United States—the Keweenaw Peninsula of Michigan, the Butte-Anaconda district in Montana, and the Ducktown district in Tennessee—and investigated the conflict between historic preservation and environmental remediation.

Landscapes

Copper mining and smelting in Montana, Tennessee, and Michigan left important legacies with which later generations would have to contend. (See table 6.1) All three mining districts were developed on important ore lodes, in stunning mountainous or hilly surroundings, near rich waterways and thick woods. All three developed or adapted technologies and social structures to overcome the challenges of the natural landscape and wrest copper from the local rock. These actions ultimately elevated two of the districts to

TABLE 6.1. Legacy and Reaction to Landscape Changes

Datum	Michigan	Tennessee	Montana
Primary Ore Type	Native	Sulfide	Sulfide
Copper Mining and Smelting Operations	1844–1969	1854–1987	1880–1980
Superfund Listing	1986	2001*	1983
Contamination Threat to Human Health	Low	High	High
Key Environmental Problem	Mill Tailings	Acid Production, Smelter Effluent	Mill Tailings, Smelter Effluent
Expected Cost of Cleanup	$15,000,000 (fifteen million)	$100,000,000s (hundreds of millions)	$1,000,000,000 (one billion)
Cleanup Funder	EPA-"Super-Fund"	Occidental Petroleum	ARCO (BP)
Post mining Economic Recovery	Heritage, University, Continued Industry	Heritage, U.S. Government	Heritage, University, Continued Industry
Heritage Priority	High	High	High
Heritage Planning	NPS-Community-NHL-HAER	Cleanup Co.-EPA-NR	Community-NHL-HAER
Community Size	Medium	Small	Medium
Earliest Museum	1920s/1958	1977	1963
Heritage Success?	High	Medium	High

*2001 is the year Occidental Petroleum agreed to clean up the Copper Basin, but to do so privately, outside of a formal Superfund designation.

top copper-producing district in the world, and the third to national importance, and created landscapes that some locals celebrated and perceived as heroic, and prompted heritage planners to preserve. Those same forces that created these heroic landscapes, however, also significantly degraded the environment. The combination of mining decline, economic decline, blight, and loss of prestige with the need to clean up past egregious behavior and the stigma of Superfund led to a contrasting landscape of defeat.

Both historic mining and environmental degradation legacies led to important decisions regarding these opposing community needs. On the one side, the desire to celebrate and commemorate the once-significant past and lure visitors to their region drove locals to preserve historic buildings and landscapes, develop museums, and try to recapture some of the district's former glory, character, and identity. The value of the tourism market, which had become the second or third leading industry in each state by the 1950s, was not lost on local heritage planners in the economically declining mining regions. On the other hand, the desire to clean up the region and eliminate the stigma of sometimes very serious contamination drove efforts to remediate the landscape as thoroughly as possible and move on to redevelopment projects not necessarily directed at short-term visitors, but instead aimed toward creating investment opportunities for people who already lived there or luring long-term investments from outside the region. Negotiating a resolution between these two competing legacies, represented in principal by the National Historic Preservation Act (NHPA) and the Comprehensive Environmental Response, Compensation, and Liability Act (CERCLA), ultimately created the landscapes of compromise.

The heroic landscape of the Butte-Anaconda district reflected its large scale. Sitting in a towering and beautiful mountainscape at the Continental Divide, the one-time largest nonferrous mining corporation in the world built the largest smelter complex and tallest smelter stack. Although Michigan mining and smelting structures were built on a smaller scale, they sat in a region of arguably equal beauty. Its one-time world-dominant copper production began with the first large-scale mining boom in the United States, was dispersed over a one hundred–mile three-county mineral range, produced 82% of the nation's copper at one time, and had some of the world's deepest hard rock mines. Although Tennessee's copper production never rivaled Michigan's or Montana's, it created, although does not celebrate, the largest sulfuric-acid-making plant in the world and perfected nearly fuel-free pyritic smelting. To the residents who live there, however, the heroic landscape (barren and treeless) reflected a very long-lived and prosperous industry, but for others served as a landscape of defeat.

The early landscapes of defeat in Montana and Tennessee reflected mostly the pollution residue from processing and smelting sulfide ores, which had created significant and early smelter smoke problems and led to important lawsuits against the mining companies. Although the Anaconda Copper Mining Company (ACM) had resolved many of the legal challenges to its operations by 1933, the Tennessee Copper Company (TCC) and Ducktown Copper Sulfur and Iron Company (which TCC bought in 1936) were, in the 1930s, just coming to terms with the scale of vegetation loss and erosion left by nineteenth-century smelting operations. Although ACM would eventually use the cost of compliance with pollution regulations to justify closing its expensive and outdated Washoe smelter in 1980, Anaconda (as an ARCO subsidiary) had largely operated without significant concerns about environmental litigation from the 1930s until the EPA came in under Superfund in the 1980s. The Tennessee Copper Company, or one of its later corporate incarnations, on the other hand, would work with federal and state agencies for seventy-five years to remedy the pollution-generated denuded landscape before it, too, faced larger remediation issues under Superfund in the 1990s, reflecting a very different experience with federal agencies than that of Montana or Michigan.

The Michigan landscape, however, never suffered from significant environmental problems in its early phase. The region's industry generated coal smoke, denuded some of the surrounding forests for mining needs, and dumped its tailings into waterways, but its mines were so spread out in the district that timber was never that far away, and its major twentieth-century mills, more centrally located in Houghton County, simply moved to nearby locales on larger or non-navigable streams or lakes when the Army Corps of Engineers complained that their mill wastes interfered with navigation.

The more serious landscape of defeat for all three districts occurred during the declining or postmining periods after most metals processing had ended and concern about contamination swelled. After a century of mining, tailing deposits in Montana were heavily contaminated with a variety of toxic and carcinogenic materials, including heavy metals, arsenic, and processing chemicals. Tailings, often dumped directly into nearby streams, were carried as far as 125 miles down the Clark Fork during spring floods, littering and contaminating its entire length from Butte to the hydroelectric dam in Milltown. Further, arsenic dust from smelting was distributed through smoke and settled on soils and in homes, creating especially dangerous conditions near smelter stacks. Acidic waters drained from the mines, and chemical residues from mining and smelting, slag, and ash were all dumped on or indiscriminately disbursed into the landscape, creating groundwater and air contamination. All of these conditions led to the largest Superfund site in the country—and one of the costliest.

Building- and community-loss furthered the sense of defeat in the Butte-Anaconda landscape. Significant losses began with the opening of the Berkeley Pit in 1955, and continued in the 1970s with the threat of pit expansion, the rash of arsons, and general neglect. The growing complexity and cost of the Superfund project, beginning in the 1980s, only exacerbated these earlier defeats. EPA planners often considered (contaminated) building demolition a cheaper alternative to remediation and preservation, as at the Washoe Smelter in Anaconda. Further aggravating these conditions were highly visible events, such as the relocation of Mill Creek residents in 1986 and the death of 342 snow geese in 1995.

Tennessee was also forced to deal with significant contamination, including large lead accumulations from its several acid plants, slag and ash dumps near former smelting sites, chemical spills, acid mine drainage, and heavy metal- and chemical-laden sediments that washed into the Tennessee River watershed from major erosion. Further, the smoke-denuded moonscape in the middle of heavily wooded East Tennessee became a national, if not international, spectacle after a two-page photo spread appeared in *National Geographic Magazine* in 1973. Adding to the defeat, several of the district's underground mines collapsed, creating areas that were (and are) inaccessible to people, and even the community's effort to preserve a small portion of the denuded landscape behind the museum failed because a mine collapse prevented continued maintenance and tree removal. While Glenn Springs Holdings (GSH), the environmental arm of the responsible party in Tennessee, negotiated to keep the EPA from listing Ducktown on the National Priorities List (NPL) to avoid, it claimed, the stigma of Superfund association, it agreed to fund the cleanup as outlined in the memorandum of understanding (MOU) cosigned by the EPA and the state of Tennessee. Further, to stay off the NPL, GSH was required to follow the same public involvement, remedial investigation, work planning, execution, and EPA-oversight procedures as if the site been listed. Despite the success of staying clear of the NPL, the region continued to carry the stigma of a distressed and degraded landscape because of persistent imagery, public involvement, and the ongoing EPA-guided cleanup.

Nonetheless, the cleanup remained privately funded and privately managed on private land, and, although Glenn Springs publically promoted its efforts to remediate the area and contributed to numerous community projects, it was still a private project and not required to abide by Section 106 of the National Historic Preservation Act as interpreted by the EPA, Tennessee SHPO, and GSH. Over the entire process from the cessation of mining, Ducktown, like Butte-Anaconda, lost much of its structural character, not only due to economic decline and neglect, but also to cleanup decisions that selected less expensive remediation alternatives to preservation.

Michigan, on the other hand, never suffered heavily from environmental problems or stigmas. The district, because of its age and disbursed mining locations, lost many of its structures to neglect, abandonment, or redevelopment, as mining and economic centers moved closer to Houghton over one hundred years. Despite this loss, the abandonment of places, in some ways, led to the growth of early heritage tourism and the exploration of ghost towns. The biggest problem the EPA identified in Michigan under Superfund was copper- and heavy metal–contaminated stamp sands only posed a human health and exposure threat if the sands were going to be redeveloped into residential property. Otherwise the sands were considered a less critical environmental threat. The EPA covered the tailings with a soil and vegetative cap, a loss to some in the district who considered the tailings an important and largely benign historic feature of the landscape.

In all three districts, many of the environmental remedies simply erased character and identity. In Montana and Michigan, regraded landscapes and smoothed-out features designed to keep new soil intact while non-native grasses grew left a muted and somewhat subdued, or even boring landscape. The revegetation efforts in Tennessee were limited to the few tree and grass species that could tolerate the contamination, leading to significant lack of initial biodiversity in the "new" woods, which feel more like a tree farm than a forest; of course, increased speciation may occur over time.

Landscapes of Compromise

In all three of the districts, the two opposing schools of landscape restoration were represented by those who wanted the landscape's natural beauty and cleanliness restored (to acceptable levels) so they could get back to the business of economic redevelopment, and those who wanted historic landscape features preserved or restored to celebrate the past and to revitalize the declining economy through tourism. However, the three regions did not hold on to their historic character equally. An important key to successful preservation negotiation was often the size and diversity of the preservation-minded population and how compelling its arguments were against certain types of remediation alternatives.

Similarities existed among the districts. Both Montana and Tennessee had large petroleum companies ultimately take responsibility for the remediation, spending from hundreds of millions of dollars to one billion dollars, some of which went to preservation projects in both districts. Both had significant air pollution problems from sulfur smoke disbursed over a wide area and significant tailings or water runoff problems affecting rivers that spread their contaminants downstream. Both had air and water problems that affected areas far away from the mining district leading to significant lawsuits. Both had large, postmining contaminant disposal problems and highly visible landscape degradation. Both the Tennessee Copper Company and Anaconda operated demonstration farms in the twentieth century in a vain attempt to prove there was no contamination.

Michigan and Butte shared other characteristics. Both regions led world production at one point. Both had relatively large populations and diverse heritage planning communities augmented by new arrivals to the region. Both had state technological universities in the mining district. Both had significant heritage projects that led to National Historic Landmark designations and HAER documentations. Tennessee and Michigan each had early mining histories impacted by the Civil War, were in naturally beautiful areas, and wished to celebrate their pasts, but had little else in common in terms of this book's discussion.

Whereas Ducktown had a relatively small, stable population of long-time residents, Butte-Anaconda had a larger population with influxes of new people at various times related to new government programs in the 1970s and the presence of Montana Tech. Butte also had a major interstate intersection and airport that easily brought people into and through town. Its larger population allowed Butte to generate several special interest groups focused on preservation. Many of its new residents were educated, and some were even interested in the preservation of industrial remains. These outsiders came to Butte-Anaconda with new perspectives and expectations to contrast the landscape with as well. Few people came from regions with landscapes emulating the Berkeley Pit, or hillsides dominated by gallows frames or structures like the Anaconda Stack, or a Superfund site as large, and the new residents evaluated the landscape as it stood when they arrived, not how it used to be.

The special interest groups and preservation-minded newcomers augmented a preservation movement begun by locals in the 1950s. Together, they generated National Historic Landmark nominations, created a museum, started an archive, and passed several local preservation ordinances that included historic mining and industrial features. They pursued many projects and, with knowledge of preservation law and well-organized efforts, were able to exert political pressure, articulate preservation desires, and negotiate long-term plans to confront the pending loss of cultural resources and historic landscapes by Superfund projects. Even the State Historic Preservation Office ceded control over Section 106 issues to the local counties (to insure speedy resolution of issues generated by the cleanup). Overall, the Butte-Anaconda district preservation population was the most organized, bottom-up-oriented, and successful of the three.

Although Butte-Anaconda had the largest Superfund site, in some ways it was the also best set up to negotiate and promote heritage—or, to corrupt a concept from M. -L. Quinn cited in Chapter 4, the region had *heritage-susceptibility*. With a sizeable population of educated, preservation-oriented, and planning-minded people aware of the law, an established National Historic Landmark district, maybe even a bit of historic activism left in the once heavily unionized community, and a very wealthy, but not necessarily cooperative, responsible party with a desire to end the expensive cleanup as soon as it could, heritage compromises were frank and decisive.

Further, the district was successful in getting National Park Service recognition through the HAER project and successive National Historic Landmark district expansions, although unsuccessful at getting a national historical park in Butte-Anaconda. Whether a formal unit from the National Park focused on historic mining and smelting would ultimately have helped the active heritage community or complicated negotiations is unknown; in the long-run, the Butte-Anaconda community was able to define what it wanted and how it wanted it—and to negotiate and compromise to get much of it.

One of the key factors influencing the Ducktown heritage process, on the other hand, was a heavy EPA and corporate presence that largely overshadowed the local preservation population, which was relatively small. Although both the EPA and Glenn Springs Holdings (GSH) considered and interacted with the locals, many of whom were not dissatisfied with GSH, the company largely drove both the remediation process and heritage process, making heritage decisions with limited public consultation. Further, the population of the region, many of who wanted heritage and the thirty-acre denuded landscape behind the museum preserved, was small. Although the small size of the population ultimately made it easier for Glenn Springs to make effective community contributions and generate local support, the population was not large or diverse enough to generate planning committees to oppose certain heritage decisions or even to lobby for different actions. Often the cleanup company talked to the same group about remediation, as mandated by the MOU, as it did about heritage issues. Further, the Ducktown Basin Museum, the key heritage organization in the area, was too underfunded and understaffed to really push for a strong heritage plan it would have to administer in the long run. Despite some well-intentioned GSH- and EPA-managed heritage projects in the overall work plan, these conditions ultimately led to a relatively weak voice for preservation, which was worsened by the accepted inapplicability of Section 106. In many ways this effort was the most top-down planning of the three districts. Moreover, heritage decisions were being made largely by remediation experts, not by heritage experts.

The key difference between the Michigan cleanup and the Ducktown and Butte-Anaconda cleanups was its simplicity. Michigan's Superfund project dealt mainly with tailings dumped and exposed in Torch Lake, Portage Lake, and Lake Superior. Outside of relatively minor removal of contaminated wastes and storage containers, the project mostly graded, capped, and covered exposed tailings at ten of the thirteen locations. This relatively small area and relatively simple work resulted in a comparatively inexpensive cleanup compared to the other two, and also likely influenced EPA's withdrawal of its normally aggressive pursuit of potentially responsible parties early in the process. Although the project was federally funded and managed to make Section 106 applicable, only one of the sites, the Quincy Smelter, had properties determined eligible for the National Register.

The Michigan Copper Country, like Butte-Anaconda, had a fairly stable population and a means of drawing new people in through a local university, which had an active program dedicated to industrial heritage that sponsored heritage research throughout the region. Further, after 1992, the new Keweenaw National Historical Park (KNHP) brought in additional educated historians and historic architects to work with the nineteen formal community-partner sites and many other related sites, and also completed regional management planning documents and cultural landscape reports that supported the NPS mission. In addition, unlike the other regions, the Keweenaw had a well-established patron in the U.S. Senate who actively pursued funding earmarks for heritage projects, including the smelter. So, in many ways, Michigan's landscape diversity extended to its preservation community, too. With community groups similar to those in Butte, Michigan's Copper Country hosted a relatively large cadre of professional heritage practitioners. Michigan copper heritage, as opposed to Montana and Tennessee, however, had neither a strictly bottom-up nor a top-down approach to heritage management.

Of the three sites, Michigan had the simplest heritage conflict with Superfund, but also the least amount of cleanup-generated money available for alternate projects like preservation. Although the remediation plan was fairly simple, the EPA encouraged preservation by recognizing the importance of the smelter (which also simplified its workload and final costs), but also complicated preservation when it attempted to extract money from NPS to pay for asbestos removal at the site, forcing the most active heritage professionals away from the project for nearly two years.

In the long run, however, the EPA worked to find compromise solutions to remediation issues at the Quincy Smelter to appease environmental concerns generated by the state of Michigan, while respecting the heritage of the site to appease the NPS, Sections 106 and 110 of the National Historic Preservation Act, and many members of the community. Overall, however, copper mining and smelting heritage in Michigan was not affected very much by Superfund. The EPA focused on stamp sands, but not all of them, and Michigan's other historic copper mining resources were spread out over a wide area, most of which the EPA was not interested in. Preservation challenges therefore had much more to do with combating neglect and inappropriate redevelopment than anything else, which is ultimately part of the KNHP's mission—not to argue that all sites in the region need preservation, which is not only not necessary, but impossible.

In the end, all three regions successfully commemorated their pasts with museums, historic sites, and National Register nominations, but they were not equal in how well their heritage was expressed in the landscapes. The two larger districts, with more significant production, elevated their sites to National Historic Landmark Districts and welcomed HAER documentation projects. All three districts had interpretive signage explaining historic processes that gave their districts significance, but Ducktown and Montana, home to more significant environmental damage and cleanup, interpreted the desecration of the environment caused by mining—and the *near-heroic* efforts to remediate it—alongside the technological and social histories. Part of the success in Michigan and Montana has to reside with the influx of newcomers interested in the historic landscape. Although many people involved in preservation in both communities were local, newcomers brought a perception and viewpoint not compromised by loss and defeat and helped local heritage preservation coalesce into a movement.

On the other side of the negotiation, the EPA's mission to protect human health and the environment drove its activities and defined how it approached other, less mission-direct activities. In Superfund projects, EPA project managers, often from district headquarters far removed from the region, approached an NPL-listed site to find a quick "fix" for the problem of contamination, without the attachment or regional identity of a local resident. While this detachment could make remediation decisions simpler and appear less biased, it also meant that managers who participated in heritage decisions were missing an important heritage value associated with local identity. EPA managers understood that the community must be involved and that the EPA must consider residents' opinions and factor those opinions into remediation decisions—especially because there are and will always be multiple ways to approach and negotiate remediation remedies. They also understand that heritage concerns must be addressed. It is in this dialogue that most heritage decisions are discussed, and the more forceful the heritage voice, the more preservation is considered.

Although, in general, its mission drives EPA's activities and decisions, and the agency does not have a strong culture of heritage-protection, some individual EPA managers at least respect a community's desire for heritage, even if they don't share it. In both Tennessee, where the EPA manager came from Atlanta and actively participated in heritage decisions with the Glenn Springs project manager, and Michigan, where EPA planners from the beginning considered the Quincy Smelter an important historic resource, individual EPA managers demonstrated compassion, and at times even enthusiasm, for some preservation, but they do not share the understanding of professional heritage planners or even heritage-oriented community members. But this consideration is not a universal trait of EPA managers, even those working the same project.

The Butte-Anaconda district was different. Because of its size, the EPA staff contingent was spread from local offices in Butte and Anaconda, to a regional office in Helena, Montana, to the district office in Denver, Colorado. Because of the size of the project, the EPA also hired local people to run the local offices, and those with a regional association and identity with the landscape identified with the strong local heritage community, and although they were still EPA mission–driven, they were able to soften the negotiations. But others and outsiders were just as intent to smooth out the contaminated bumps in the landscape and to remove blight to foster a simple remediation process.

Resolving a Heritage-Remediation Conflict

Heritage preservation poses potential conflicts in a variety of circumstances, from large environmental remediation projects to small municipal redevelopment projects. By law, however, heritage must be considered whenever the federal government, federal funds, federal permits, or federal licenses are needed and the National Historic Preservation Act covers a wide number of projects at all levels of government. As we have seen in this study, because of different historical, demographic, economic, geographical, and political circumstances, heritage-remediation conflicts were resolved in varying manners in the Michigan, Tennessee, and Montana copper districts, and heritage proponents enjoyed varying degrees of success. What, then, are the most basic lessons we can take away from this study of heritage-remediation conflicts in three nationally important copper mining and smelting districts that have broad applicability to other types of heritage conflict; and, in the face of often-aggressive remediation/redevelopment efforts, what steps can those who wish to preserve the heritage of their regions take to ensure maximum preservation? Four keys to success in promoting heritage emerged from these three case studies: organization, knowledge, persistence, and political sympathy or support from key state agencies or political patrons in Congress.

Organization is the first key. As our three case studies have shown, the better the preservation elements in a community are organized, the stronger their message and political influence becomes, ultimately resulting in a social force that must be contended with in a much more formal way by remediation-minded authorities than a small group or individuals would be. The successful organization of grassroots groups in Butte-Anaconda illustrates this notion directly, and the affiliated heritage partners of the Keweenaw National Historical Park, a large, even loosely, coordinated group, have provided the old Michigan copper district with preservation advocates capable of encouraging remediation and development advocates to negotiate and compromise on heritage issues, rather than simply bulldoze everything in its way.

Second, knowledge of the Comprehensive Environmental Response, Compensation, and Liability Act (CERCLA, Superfund) and other applicable statutes is critical. The laws normally used to initiate and coordinate environmental remediation specifically state that project planners consult with local communities and consider other relevant legislation, such as the National Historic Preservation Act (NHPA) and other preservation laws. This knowledge allows an organized group to make strong legal arguments, because neither CERCLA nor NHPA refutes the other, and on-site conflicts usually arise out of the inconsistent interpretation of standards outlined in the laws or out of ignorance of their intended aims. Further, knowledge of successful alternative approaches to similar remediation problems gives decision makers more options, as in the example of the Old Works Golf Course slag-based sand traps in Montana. Knowledge that smelter slag was being safely used in direct contact with golfers was cited by heritage advocates in Michigan to influence slag heap preservation at the Quincy site. Of course, recruiting a local attorney interested in heritage preservation to work, hopefully gratis, with local heritage proponents would also contribute to the local knowledge base on legal issues and recourses.

Third, hedging against heritage loss requires persistence. Butte-Anaconda and the Keweenaw each pursued heritage planning over decades. Butte's first heritage plan began in the 1950s, and continued with successive plans and expanded planning efforts in the 1960s, 1970s, 1980s, 1990s, and its latest plan and National Historic Landmark district designation were finalized in 2005. The Keweenaw, too, started with preservation projects in the 1950s and, in 2009, was working on its latest National Park Service–sponsored regional cultural management plan.

Lastly, heritage preservation and environmental remediation are both, ultimately, political in nature. Recognition of this fact, and taking the appropriate steps to gain political support, can be critical to the level of success heritage proponents can secure in preservation-remediation conflicts. Of course, the previous keys—organization, knowledge, and persistence—can contribute significantly to the ability of local heritage advocates to secure such political support, too. At the state level, securing the active support of key state agencies, especially the State Historic Preservation Office (SHPO), the agency specifically charged by the NHPA with oversight of cultural resources, can lend significant negotiating weight to heritage proponents in conflict with remediation agencies. For example, in Butte-Anaconda, the Montana SHPO enhanced local heritage planning and preservation efforts, whereas the weak involvement of the Tennessee SHPO all but sealed the fate of important cultural resources there. At the national level, too, securing political patronage or support is important, especially because the primary remediating agency—the Environmental Protection Agency (EPA)—is federal. A political patron at the national level can help reign in an overly aggressive EPA, secure funding to enhance heritage preservation, or help secure formal national-level recognition of "historical significance," such as Michigan senators did in forming the Keweenaw National Park and in identifying earmarks for the Quincy Smelter, and Montana's senators did in the establishing of high-tech projects in Butte.

All four of the keys outlined above—organization, knowledge, persistence, and political support—are important. Together, they give heritage proponents significant clout and encourage remediation authorities to seek compromise, because, in the long run, compromises are cheaper than prolonged debates and court battles.

Appendix

Mining District Heritage Model

The question posed in the introduction of this book asked how heritage-minded groups, living in historic industrial communities and wishing to commemorate their past and preserve their historic resources articulate and negotiate that desire in the face of sometimes overwhelming opposition. The answer required an understanding of the economic, demographic, and value systems of the communities being examined. The outcome of this analysis suggested a generalized model for the experiences of the three districts, which could be extrapolated to examine other rural mining and smelting locations facing declining fortunes and increasing concern about the legacy of industry.

Each of the three districts went through five distinct phases: (1) significant growth in the nineteenth century, (2) periods of high production in the early twentieth century, (3) a period of decline that ultimately led to (4) the abandonment of most mining and smelting operations by the end of the twentieth century. All three districts also had significant and persistent environmental issues leading to (5), a listing on the Superfund National Priorities List (NPL) or, in the case of Ducktown, Tennessee, a negotiated non-listing with most of the legal and procedural requirements of a listed site. All three of the districts also spawned significant and expanding heritage movements to commemorate their once-proud histories—movements that began as the output of each district began to decline. At each phase—growth, high-production, decline, abandonment, and Superfund—issues of economics and environment/landscape change had clear implications for the identity and external perceptions of each region, and many of these issues unfolded along parallel lines. This appendix will explore a model that illustrates how historic districts contend with both their contaminated pasts and heritage concerns and may have implications for future studies of related historic industrial districts undergoing remediation. Note: The following graphs are general representations of historical data, not quantitative representations of scientific data. They reflect sentiments expressed in period literature and are supported in the text with examples.

A simplified life span of an extractive industrial district can be described through a rough bell curve based on the productivity generated from its mines (see Figure A.1). In most instances, successful mines have significant growth in their early years, leading to varied production highs; but as ore quality and quantity dwindle, mining declines. Some districts may have substantial deposits of lower-grade ores that, after the richer early ores are mined, can prolong mining, but these new ores usually require a significant shift in the manner and scale of operations.[1] Ultimately, however, when the quality of ore gets too low to mine and process profitably, the district's extractive history comes to an end, often leaving behind physical remnants of the once-productive landscape including structures, mining cavities, and mining and processing wastes accumulated over decades of activity with nonexistent or, at best, minimal waste-handling regulations. Montana copper production, which, for example, did not exist in any significant quantity before 1880, grew to produce 100 million pounds per year in 1887, outproducing the reigning Michigan district. By the late mid-1910s, Montana was

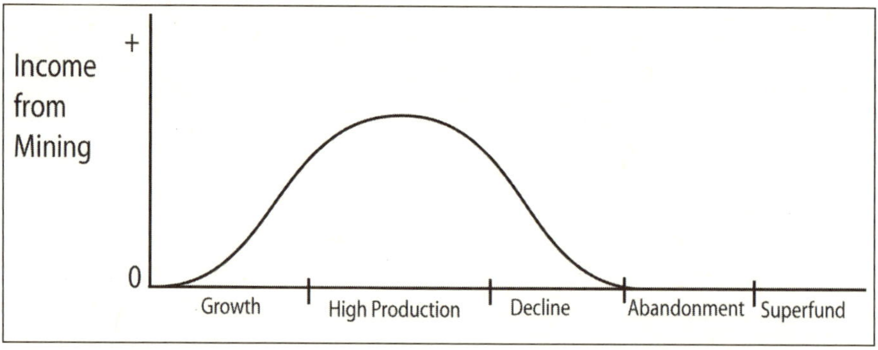

Figure A.1. Income from mining, broadly defined and based on productivity, increases through growth and high-production periods then declines as ore quality decreases to a point that mining becomes unprofitable. The growth period in most profitable and long-lived districts, however, is relatively short and steep and the decline is long and slow. By the author.

averaging over 300 million pounds of copper per year before declining back to 100 million pounds in the mid-1940s. Although the Berkeley and later Continental pits sustained production in the district through this writing, underground mining came to a halt in the 1960s.

The population size of a successful district follows a similar trajectory to income and productivity, but tends to stabilize after mining ends (see Figure A.2). In many cases in the United States, nineteenth- and twentieth-century mining deposits were found in rural and isolated areas as geologists ventured west exploring new territories. Often after the removal of indigenous populations, mining and metal processing expanded rapidly during the growth and high-production mining stages with sometimes vast numbers of European and Euro-American settlers moving in and building towns with downtown districts, theaters, saloons, and new housing developments to support and service the burgeoning new populations. However, when mine production slowed, leading eventually to operational abandonment, many people left the regions looking for work. They left behind a much smaller group of people, now many second-, third-, or fourth-generation residents, who considered themselves locals with a birthright to the heritage of the region and a strong affinity for the landscape. They tried to sustain their communities in the face of declining economic conditions. Members of this group were often the initiators and supporters of the first heritage and commemoration projects—and also the key population the EPA was concerned about protecting. Houghton County's population, for example, increased during the region's mining growth period, reaching a recorded high in 1910 of 88,098. Following peak production later in the decade, the county's population dropped to 71,930 in 1920, then to 52,581 in 1930. By the 1950s the population reached the mid-30,000s and has remained near 35,000 from 1960–2000.[2]

In the growth and high-production mining phases, regional identity was generally favorable and based on the success of mineral operations and the importance of its mines, their financial returns, and their associated political power. The region itself, if not the state within which the mines, mills, and smelters resided, sometimes had high average individual incomes and a strong tax base directly derived from mining. The mineral companies themselves often had fairly strong political influence on local and sometimes

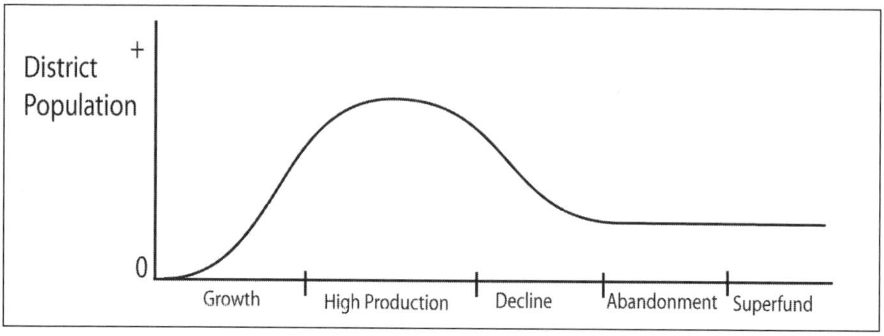

Figure A.2. District population of mining and metal processing regions will increase through growth and high-production phases, then as mining declines, many people will leave the area searching for work, ultimately leaving a relatively stable population of long-term residents. By the author.

state and national governments, and were thus able to keep their operations free from many economically distracting influences, like foreign competition, waste disposal restrictions, and organized labor. Most local residents in these periods of growth were not really local, but new to the region, lured by the prospect of well-paying jobs, a larger affiliation with the industry and favorable views toward the work[3] (See Figure A.3). When the Anaconda Copper Mining Company beat Calumet and Hecla's attempt to drive down prices and force the Montana industry out of business, and Butte became the largest copper mining district in the world in 1887, local newspapers proclaimed the victory. Similarly, during their periods of high production both Michigan and Montana copper districts celebrated their contributions to copper mining proclaiming their *world-leading* output and the *world's largest* steam-hoist, smoke-stack, and smelter, and the *world's deepest* copper mines.

However, as production slows and infrastructure improvements decline, less investment goes into the workforce often causing strikes over unsafe working conditions or fears of unprosecuted arsons; and, with significant environmental decay, morale and external perception of the once-productive district falls, sometimes to very low levels. Anacondans lamented after the closure of the smelter that "they cut off our primary reason for existence,"[4] and Ducktown's community believed there was a strong stigma associated with a Superfund listing. Further, Butte and Anaconda's designation as the "ugliest spot in Montana" and one of the "world's must-miss vacation

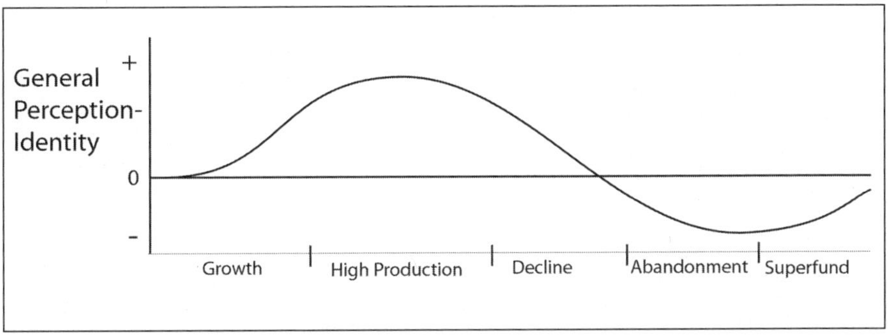

Figure A.3. General perceptions and identity of a region are usually positive as the district is producing wealth, but decline as output slows, incomes drop, and the reality of environmental contamination set in. It will likely rise again as remediation efforts finish and the region is officially delisted from the NPL and heritage projects commemorate the region. By the author.

spots," and articles on Tennessee's Copper Basin titled "the Death of Ducktown" and "Copper Basin Cover-up" clearly echo the negative perceptions generated by outsiders during periods of decline. With positive redevelopment in these regions—both in terms of growing heritage consciousness and remediation of obvious health risks—perceptions of life in the region, both internal and external improve.

While production increases through the first two mining phases, the landscape changes accordingly, and sometimes at alarming rates, often leading to some awareness of growing environmental threats (see Figure A.4). The wastes generated by mining operations were only important to the mining companies if they presented a production bottleneck problem or could, at some point in the future, be reprocessed for income-generating by-products or reprocessed for valuable materials lost using earlier, less efficient concentrating methods.

Actual environmental degradation during the first phases, however, was only important to those who were not part of the capital-generating machine or who were outside of the company influence and being negatively affected. These parties usually had limited recourse except to take the offending industries to court and sue for property damages. In some instances, the offending industry simply paid the damages, acquired the damaged properties, or installed pollution-control equipment if the toxic waste had some salable value. These options were often considered a less costly alternative to

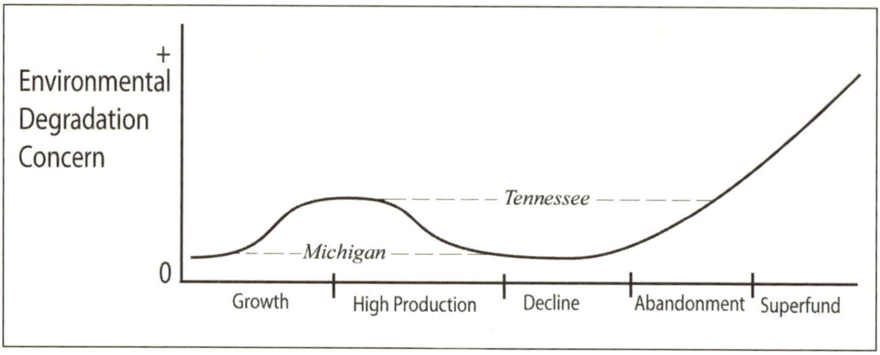

Figure A.4. Environmental degradation concern, reflected in costs spent responding to environmental problems, has a modest effect on operations in the growth and high-production phase, if at all, that declines if the mine has enough capital and political influence to buy out those concerned. However, concern increases considerably as mining slows and communities assess the long-term environmental costs of having hosted extractive industries that may no longer provide any economic benefits to the area. However, in some districts, growth-period pollution did not exist on a large enough scale to warrant early concern like Michigan, whereas in other districts, the early pollution concern did not decline, and pressure remained high until later environmental concern mounted in the abandonment and Superfund periods like in Tennessee. By the author.

fighting court cases, even though a mining company would likely win. If the environmental landscape change was significant enough and highly visible, then the problems were written about in newspapers, journals, or textbooks, but still little was usually done to hamper the income-generating capacity of the industry. Georgia and Montana farmers and landholders, for example, suffered from the deleterious effects of mining and smelting and took the offending companies to court in the early twentieth century. Anaconda successfully handled its pollution problems with settlements, land purchases, and negotiated changes to practice, and, after environmental degradation concerns dropped in the early 1930s, operated largely without concern for pollution until the 1970s. Although the Keweenaw never saw an early concern for environmental contamination, the Ducktown district never lost the stigma of pollution. After lawsuits forced them to change practices—albeit to new systems that resulted in significantly improved profits—Tennessee copper companies began remediating the landscape and continued to do so in 2009. Its concern never went away, but did increase substantially, like Michigan's and Montana's, as new toxins were identified in the later abandonment and Superfund periods. Such early concerns about environmental change form the roots of later responses to landscape remediation in the districts, especially after mining and processing slow down or are abandoned. Quieted sometimes, but never extinguished, environmental concern grows significantly after mine productivity declines, often following, but not necessarily related to, the establishment of national social-political conservation movements (especially following EPA investigations of sites under Superfund).

In the period of decline, as extractive output slows, people and communities do not earn as much as they did before, workers benefits and jobs are cut, companies do not invest in infrastructure or productive improvements as much, and the regional population starts to drop almost as fast as it grew in the early stages. With mine output and community wealth declining, which becomes especially noticeable after another mining region surpasses its once-significant productive capacity, a community identity starts to change, and second- third- and fourth-generation locals begin to notice how much money has left the region, and how much they will soon have to contend with rising unemployment, a declining tax base, declining property values, an aging population, and the environmental and health impacts left behind. With declining tax income and lessening corporate support for community projects, towns start looking worse as structures are torn down, burned down, or just abandoned and left to deteriorate, affecting the self-identity and external perception of the region (see Figure A.3).

Liability is hard to assign by these communities unless a "new" corporate entity, such as a petroleum company, can be targeted for causing much of the economic decline and contamination, as in Montana. But the new corporate entity may also be seen as the savior fixing the landscape problems for which no one in particular is to blame, such as in Tennessee. In other instances, the communities may tacitly blame the U.S. Government government for its wanton disregard for the region's well-being during the industrial decline and the negative changes in public opinion after the EPA declares an area contaminated and in need of significant cleanup. The government, in turn, often desperately searches for someone to label as responsible for the environmental problems and force them to pay for the clean up, especially if there are acute threats to human health and expensive remediation alternatives.

Despite varying accusations of culpability among stakeholders, locals often in this period start reflecting on their times of greatest prosperity and economic significance and begin commemorating their past with small heritage projects including museums, National Register nominations, and interpretive/educational programs in an attempt to hold on to—or regain—some bit of the dignity lost with the decline (see Figure A.5). Like the early concern for environmental degradation, these early heritage projects often serve as the roots for later heritage projects, especially after Superfund statues allocate funding beyond remediation and other federal laws mandate consideration of the effects of all federal projects on cultural resources. Both early heritage concerns and early environmental concerns reflect the desires of the community, but each needs a powerful agent of change like Superfund to awaken the broader community beyond the early proponents and drive the formalization of localized planning.

In the 1960s, genuine and mounting concern about the health effects of the more invisible toxins generated by industry grew following new studies on hazardous wastes and human health. This concern contributed to a growing social-political movement against environmental degradation and led Congress to enact its first statues regulating industrial effluent levels. In the ensuing decades, Congress passed specific laws limiting air pollution and water pollution, and regulating transportation, disposal of hazardous wastes, and the cleanup of abandoned hazardous waste. Most of these new laws were enacted after a century of American industrial dominance had left significant scars on the landscape, and other nations—with lower production costs and weaker environmental laws—came to dominate global mineral production.

Near the end of production in a mining district, the environmental effects of the formerly grand industry seem much more pronounced because they are no longer part of a productive income-generating process, thus popular opinion turns decidedly antagonistic. Community identity is nearly at its all-time low as locals feel abandoned and contaminated by the industry that once gave their community purpose. Almost worse, govern-

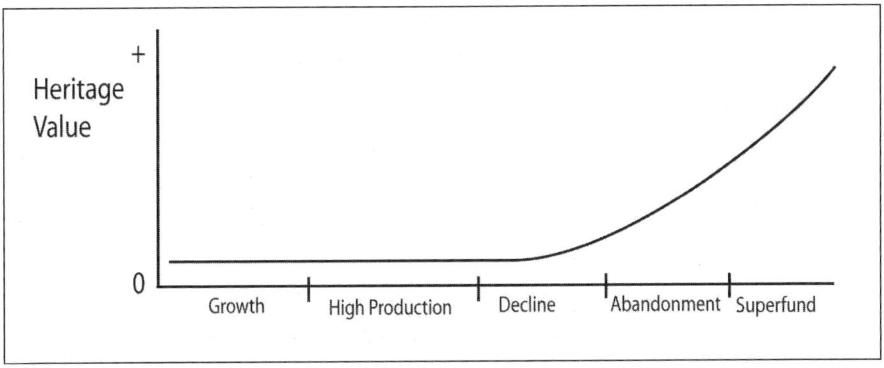

Figure A.5. Heritage values of a region, or the concern for commemorating the past, begin to grow as industries decline. Residents look to commemorate their once-glorious past as a means of retaining self-respect and generating tourism and economic development opportunities. Value and perception of the landscape become much more important as mining and mining income dwindle in the decline phase. By the author.

mental outsiders judge the town's environmental problems based on new environmental laws. If these entities decide the problems require a significant federal effort to remediate, the decision often undermines hopes of new tourism-based development. At that point, the region has been completely cleared of its mineral wealth, the community's tax base is gone, many of the jobs/people are gone, too, and all that remains are the toxic residues and stigma of being declared a Superfund site (see Figure A.3).

External judgment and the remediation process present conflicting alternatives to communities. Whereas a Superfund project in a town reflects an acknowledgment of—and remedy to—past sins, leading, hopefully, to a "clean bill of health" and a readiness to redevelop damaged lands, the very same clean up announces to the world that this place is contaminated and not suitable for immediate redevelopment. Tourism, in many ways, fits nicely into this scenario, because people visit the region for a short time, not long enough to be contaminated, but long enough to inject money into the local economy. Ironically, remediation efforts often spur historic preservation and heritage awareness because they threaten to destroy the remaining potential sources of community pride and revenue.

By the time remediation planning begins, preserving and promoting heritage become increasingly effective because parts of the cleanup legislation allow for cultural improvement; however, many of the cleanup professionals and managers, without the sense of identity shared by the locals, are mostly interested in fixing the threats to human health and the environment, not heritage preservation. The first and often least expensive remediation considerations are often to demolish standing features, and grade and cover contamination—even historically significant structures—often causing heritage-minded locals, but not necessarily development-minded locals, to feel the impending loss of their heritage landscapes and identity more keenly; as a result, they may organize to force cleanup monitors and planners to abide by the heritage laws passed to protect that very same landscape (see Figure A.5).

Notes

Introduction

1. David Lowenthal, "Stewarding the Past in a Perplexing Present," in *Values and Heritage Conservation*, ed. Randall Mason, Erica Avrami, and Marta de la Torre (Los Angeles: The Getty Conservation Institute, 2000), 18.

2. Peter Howard, *Heritage: Management, Interpretation, Identity* (New York: Continuum, 2003), 6.

3. Australia State Party, *The Burra Charter* (International Charter for the Conservation and Restoration of Monuments and Sites [ICOMOS], 1999), 2.

4. Among theories of technological change and evolution, social construction and network or systems theory largely dispel the notion that significant inventions or events had a single route of development. Rather, incremental steps are largely negotiated, contested, and influenced by a wide variety of direct and peripheral actors. See, for example, Wiebe E. Bijker, Thomas P. Hughes, and Trevor Pinch, eds., *The Social Construction of Technological Systems: New Directions in the Sociology and History of Technology* (Cambridge: MIT Press, 1987).

5. See www.nps.gov/history/nr/ accessed 16 December 2011.

6. If a privately owned site on the National Register is altered or changed so much that it no longer meets the requirements for listing, the only penalty is delisting.

7. "National Historic Preservation Act of 1966," in *Public Law 102–575* (1966) including its 1980 and 1992 amendments.

8. "Cleanup Process," U.S. EPA, accessed July 11, 2009. http://www.epa.gov/superfund/cleanup/index.htm

9. "Who Are We and How Are We Organized?" accessed July 11, 2009. http://www.epa.gov/epahome/aboutepa.htm

10. "Historic and Archaeological Resource Protection for USEPA Personnel," ed., Office of Enforcement and Compliance Assurance (Washington, DC: U.S. EPA, 2007), 1–6. Reviewed in 2008.

11. See, Thomas King, *Cultural Resource Laws and Practice* (Walnut Creek, CA: Altamira Press, 2008) and "CERCLA and NPA Coordination for Superfund Sites," NPI, accessed July 11, 2009. http://www.npi.org/sem-cercla.html

Chapter 1

1. Michael Rice Jones, "Oxhide Ingots, Copper Production, and the Mediterranean Trade in Copper and Other Materials in the Bronze Age" (master's thesis, Texas A&M, 2007), 128.

2. Paul T. Craddock, *Early Metal Mining and Production (*Washington, DC: Smithsonian Institution Press, 1995), 194.

3. R. A. Davis Jr. and others, "Rio Tinto Esturary (Spain): 5000 Years of Pollution," *Environmental Geology* 39, no. 10 (2000): 1115.

4. See Andreal Manasse and Marcello Mellini, "Chemical and Textural Characterisation of Medieval Slags from the Massa Marittima Smelting Sites (Tuscany, Italy)," *Journal of Cultural Heritage* 3, (2002): 187–206; J. P. Grattan, D. D. Gilbertson, and C. O. Hunt, "The Local and Global Dimensions of Metalliferous Pollution Derived from a Reconstruction of an Eight Thousand Year Record of Copper Smelting and Mining at a Desert-Mountain Frontier in Southern Jordan," *Journal of Archaeological Science* 34, no. 1 (2007): 83–110; F. B. Pyatt and others, "The Heavy Metal Content of Skeletons from an Ancient Metalliferous Polluted Area in Southern Jordan with Particular Reference to Bioaccumulation and Human Health," *Ecotoxicology and Environmental Safety* 60, no. 3 (2005): 295–300; and F. B. Pyatt, "Copper and Lead Bioaccumulation by Acacia Retinoides and Eucalyptus Torquata in Sites Contaminated as a Consequence of Extensive Ancient Mining Activities in Cyprus," *Ecotoxicology and Environmental Safety* 50, no. 1 (2001): 60–64.

5. Sungmin Hong and others, "A Reconstruction of Changes in Copper Production and Copper Emissions to the Atmosphere During the Past 7000 Years," *The Science of the Total Environment* 188, no. 2–3 (1996): 248.

6. See F. B. Pyatt, "Copper and Lead Bioaccumulation," Ecotoxicology and Environmental Safety, 50, no. 1 (2001): 60–64.

7. Quoted in Jerome O. Nriagu, "Global Metal Pollution: Poisoning the Biosphere," *Environment* 32, no. 7 (1990): 8.

8. E. Borsos and others, "Anthropogenic Air Pollution in Ancient Times," *ACTA Climatologica Et Chorologica* (Universitatis Szegediensis, Szeged, Hungary: Universitatis Szegediensis, 2003), 8.

9. ICOMOS, "Evaluations of Cultural Properties," ed., Bureau of Cultural Properties World Heritage Convention (UNESCO, 2001), 96–97.

10. Quoted in Anna S. Ek and others, "Environmental Effects of One Thousand Years of Copper Production at Falun, Central Sweden," *Ambio* 30, no. 2 (2001): 96.

11. Ibid., 98.

12. Ibid., 97. Falun Smelters released an estimated 40,000 tons of SO_2/year during peak production years from 1620–1640. The authors, however, caution that these numbers represent less than 10% of national SO_2 peaks in the 1970s.

13. Swansea, *A Short History of the Hafod Works* (Swansea, Wales: Welsh Assembly Government, 2007), 4 (insert).

14. Ibid., 4 (insert).

15. Charles Darwin, *Journal of Researches into the Natural History and Geology of the Countries Visited During the Voyage Round the World of H.M.S. Beagle*, 11th ed. (London: John Murray, Albemarle Street, 1913), 276.

16. Aaron Snowden Piggot, *The Chemistry and Metallurgy of Copper, Including a Description of the Principal Copper Mines of the United States and Other Countries, the Art of Mining and Preparing Ores for Market, and the Various Processes of Copper Smelting* (Philadelphia: Lindsay & Blakiston, 1858), 297.

17. Edmund Newell, "Atmospheric Pollution and the British Copper Industry, 1690–1920," *Technology and Culture* 38, no. 3 (1997): 660.

18. Quoted in Noga Morag-Levine, *Chasing the Wind: Regulating Air Pollution in the Common Law State* (Princeton, N.J.: Princeton University Press, 2003), 51.

19. Newell, "Atmospheric Pollution," 678.

20. Margaret E. Mattson and Tee L. Guidotti, "Health Risks Associated with Residence near a Primary Copper Smelter: A Preliminary Report," *American Journal of Industrial Medicine* 1 (1980): 365.

21. Piggot, *The Chemistry and Metallurgy of Copper,* 297. Land purchases were likely later nineteenth-century solutions and a practice American smelter companies would employ to avoid smoke and pollution lawsuits in the days before the Clean Air Act.

22. Morag-Levine, *Chasing the Wind,* 39.

23. Quoted in Ibid., 41.

24. Edmund Newell and Simon Watts, "The Environmental Impact of Industrialization in South Wales in the Nineteenth Century: 'Copper Smoke' and the Llanell," *Environment and History* 2, no. 3 (1996): 312.

25. Newell, "Atmospheric Pollution," 670–671.

26. Morag-Levine, *Chasing the Wind,* 50.

27. Newell, "Atmospheric Pollution," 672.

28. Morag-Levine, *Chasing the Wind,* 50–51. Although Swansea did not do as well in the 1848–49 cholera epidemic, it did fare much better than many of its contemporary

cities. See E. Ashworth Underwood, "The History of Cholera in Great Britain," *Proceedings of the Royal Society of Medicine*, XLI no. 165 (1947): 169.

29. Newell, "Atmospheric Pollution," 680.

30. John Percy, *Metallurgy. The Art of Extracting Metals from Their Ores* (London: J. Murray, 1861), 324.

31. Cited in Ibid., 339, as an example of delusional thinking regarding smelter smoke.

32. Ibid., 335.

33. Gerald Gabb, *Lower Swansea Valley Factsheet 7: The Rise and Fall of the Copper Industry* (Swansea: Swansea Museum's Service, 1990), 5.

34. Quoted in Newell, "Atmospheric Pollution," 667.

35. Ibid., 671.

36. Percy, *Metallurgy*, 339.

37. Ibid., 339–40.

38. Newell, "Atmospheric Pollution," 664.

39. Joan Day and R. F. Tylecote, *The Industrial Revolution in Metals* (London: The Institute of Metals, 1991), 155. Although the Macclesfield smoke case (1770) and the sulfur-recovery process (1778) have been noted by several scholars working with primary materials, none have linked the two. The sulfur recovery process was located at a mine site, some distance from Liverpool, but it is hard to imagine that smelting a desulfurized ore, thereby reducing the effects of smelter smoke, would not weigh heavily on a decision to partner in a sulfur recovery process.

40. Morag-Levine, *Chasing the Wind*, 48. Sulfur trioxides, on the other hand, easily combined with water to form acid. Creating SO_3 from SO_2 became the first technological hurdle to acid manufacture from smelter smoke.

41. Quoted in Newell and Watts, "The Environmental Impact of Industrialization in South Wales in the Nineteenth Century," 313.

42. Ibid., 315.

43. Newell, "Atmospheric Pollution," 678.

Chapter 2

1. R. F. Tylecote, *A History of Metallurgy*, 2nd. ed. (The Institute of Materials, 1992), 150.

2. Edmund Newell, "Copperopolis: The Rise and Fall of the Copper Industry in Swansea District," *Business History* 32, no. 3 (1990): 91.

3. Stephen Hughes, *Copperopolis: Landscapes of the Early Industrial Period in Swansea* (Aberystwyth, Ceredigion: Royal Commission on the Ancient and Historical Monuments of Wales, 2000), 63. According to Hughes, between 1860 and 1914, eleven of the fourteen smelters that stopped producing copper in Swansea were converted to zinc smelters until stiff competition from modernized American, Canadian, and Australian zinc works forced the closure of all but one of them by 1926. The one remaining zinc works, modernized with government funds, continued until 1960.

4. Larry Lankton, *Cradle to Grave: Life, Work, and Death at the Lake Superior Copper Mines* (New York: Oxford University Press, 1991), 62. The chief adaptation to Welsh reverberatory furnaces built to smelt Michigan mass copper was a door in the top to load large pieces. Whereas this addition could be considered the first step in the Americanization of smelting because it was a significant modification to a known practice, it did not increase the capacity of the furnace, was not replicated in other districts, and merely simplified the act of melting large pieces of copper.

5. Gary W. Van Lingen, "The Interpretation of Archaeological Remains at the Pittsburgh and Boston Copper Harbor Copper Mining Company's Second Campsite: Evaluating a Structural Component and Its Potential Uses and Contexts" (master's thesis, Michigan Technological University, 2002), 160–61.

6. Blast, shaft, and cupola furnaces each operated on the same principal of air forced or "blasted" into the hearth of a firebrick lined furnace stack. Crews charged the furnaces through a door higher in the stack with copper ores or mattes; fuels of coke, coal, or charcoal; and often a flux. The blasted air provided plenty of oxygen to the burning fuel, creating temperatures high enough to melt the metal, waste rock, and flux, which were drawn off through tap holes near the bottom. *Blast furnaces* were generally fixed to their location on a substantial foundation with no access to the furnace bottom except through the furnace top. *Shaft furnaces* were the same as blast furnaces, but the term was used initially to describe an Agricola top-loaded blast furnace, compared to an ancient mound furnace, and later to describe cylindrical furnaces when the shape of the blast furnace became much larger more rectangular in the late nineteenth century. *Cupolas* for smelting copper were initially small cylindrical blast furnaces on pedestal legs with removable bottoms. They were largely portable, although by the twentieth century, they resembled large rectangular blast furnaces with access to the furnace bottom from below. Despite these technical differences, the literature from the mid-nineteenth through the twentieth century often uses the three terms indiscriminately.

7. Thomas Egleston, "Copper Refining in the United States," *Transactions of the American Institute of Mining Engineers* 9 (1881): 683; Frederick Laist, "Seventy-Five Years of Progress in Smelting and Leaching of Ores," in *Seventy-Five Years of Progress in the Mineral Industry, 1871–1946* (New York: The American Institute of Mining and Metallurgical Engineers, 1948), 129–31; Charles Kuzell, "The Development of Modern Copper Smelting," *Transactions of the Metallurgical Society of AIME* 218 (August 1960): 582; Allison Butts, *Copper: The Science and Technology of the Metal, Its Alloys, and Compounds* (New York: Hafner Publishing, 1970), 89.

8. H. O. Hofman, *Metallurgy of Copper* (New York: McGraw-Hill, 1924), 349.

9. See, for example, Lankton, *Cradle to Grave*, 12.

10. William Bryam Gates, *Michigan Copper and Boston Dollars: An Economic History of the Michigan Copper Mining Industry*, Studies in Economic History (Cambridge: Harvard University Press, 1951), 4, 202, 216.

11. Larry D. Lankton, *Beyond the Boundaries: Life and Landscape at the Lake Superior Copper Mines, 1840–1875.* (New York: Oxford University Press, 1997), 119.

12. James B. Cooper, "Historical Sketch of Smelting and Refining Lake Copper," *Proceedings of the Lake Superior Mining Institute* 7 (1901): 44; Charles T. Jackson, "Geological and Mineralogical Report," ed., United States Senate (Washington, DC: 1849): 443–444; Patrick Martin, *Technical Report on the Archeological Survey of Isle Royal National Park* (Houghton: Michigan Technological University, 1986), 15; and Donald Chaput, *The Cliff; America's First Great Copper Mine (Grand Rapids, MI: Sequoia Press)*, 18.

13. Egleston, "Copper Refining in the United States," 679.

14. Ibid., 680.

15. Gates, *Michigan Copper and Boston Dollars*, 28.

16. Ibid., 681.

17. Egleston, "Copper Refining in the United States," 681–682.

18. The first charge into the smelting furnace in Wales usually required a copper content of less than 10% to ensure a considerable slag to remove iron without much copper oxide formation. Because of this need, a lot of fuel was spent heating and melting 90% waste. With the cupola process, a richer charge could be used in the first fusion in the reverberatory furnace, and the resulting copper-rich slag could be smelted in the cupola. This process proved so successful that John Vivian patented the process in England in1859. (See Chapter 3)

19. Charles K. Hyde, *Copper for America: The United States Copper Industry from Colonial Times to the 1990s* (Tucson: University of Arizona Press, 1998), 24, and

Christine E. Davis, *Archaeological Survey and Land-Use History of The Pittsburgh Technological Center Site, Pittsburgh, Pennsylvania* (Survey Report, Pittsburgh: Urban Redevelopment Authority of Pittsburgh, 1989), 26–27.

20. "The Detroit and Lake Superior Copper Co.'s Smelting Works," *Detroit Post and Tribune Newspaper* 1879, Supplement, 1.

21. Cooper, "Historical Sketch of Smelting and Refining Lake Copper," 45–46.

22. Henry D. Conant, "Copper Smelting in Michigan," *The School of Mines Quarterly* 42, no. 4 (1911): 286.

23. Ibid., 682.

24. Silas Farmer, *The History of Detroit and Michigan; or, the Metropolis Illustrated; a Full Record of Territorial Days in Michigan, and the Annals of Wayne County*, 2nd ed. (Detroit, MI: S. Farmer & co., 1889), 818.

25. Ibid., 197. The first locks at Sault Ste. Marie opened in 1855, allowing a continuous passage from Lake Superior to destinations on the lower lake and significantly reducing transportation costs. Further, sandbars that had blocked the entrance to Portage Lake were dredged by 1860, clearing the way for navigation all the way to Houghton and Hancock.

26. Cooper, "Historical Sketch of Smelting and Refining Lake Copper," 46.

27. Gates, *Michigan Copper and Boston Dollar*, 43.

28. Egleston, "Copper Refining in the United States," 685.

29. Ibid., 686.

30. Henry D. Conant, "The Historical Development of Smelting and Refining Native Copper," *Mining Congress Journal* 17 (1931): 532.

31. Lankton, *Cradle to Grave*, 12, and Laist, "Seventy-Five Years of Progress in Smelting and Leaching of Ores," 127. Compared to other types of copper smelting this number is fairly high. Welsh sulfide smelting required a higher waste content to maximize slag to remove iron and minimize copper loss. Smelters generally wanted less than 10% copper content entering the furnace.

32. Ibid., 897–927.

33. Hyde, *Copper for America*, 54–55; Gates, *Michigan Copper and Boston Dollars*, 73.

34. Conant, "The Historical Development of Smelting and Refining Native Copper," 532.

35. Titus Ulke, "Notes on the History of Electrolytic Copper Refining in America," *Electrochemical Industry* 1, no. 7 (1902–3): 240.

36. C. P. Wilson, "A Brief History of Copper Smelting on Portage Lake," (unpublished graduate paper available at Michigan Technological University Archives, Michigan Technological University, 2008), 4–6.

37. Larry D. Lankton and Charles K. Hyde, *Old Reliable: An Illustrated History of the Quincy Mining Company* (Hancock, MI: Quincy Mine Hoist Association, 1982), 152–53.

38. Patrick Martin and Gianfranco Archimede, *The Quincy Mining Company Smelting Works, 1898: Historical Land Use Survey Project* (Houghton: Michigan Technological University 2002), 3.

39. Charles K. Hyde, "Historic Report on the Quincy Mining Company," (Washington, D.C.: National Park Service, 1978), 174.

40. Martin and Archimede, "The Quincy Mining Company Smelting Works," 11.

41. Trepal, "The Calumet & Hecla Smelting Works: 1887–2006" (unpublished graduate paper available at Michigan Technological University Archives, Houghton, MI), 15. The steam turbines powered by waste heat boilers provided enough electricity for C&H's entire smelter plant needs, and Quincy's plants generated enough electricity not only to power the smelter, but also to send additional electricity up to its mines, offices, and residences.

42. Ibid., 7.

43. Gates, *Michigan Copper and Boston Dollars*, 199–200. The EPA reported an estimated 200 million tons of tailings dumped into Torch Lake alone from the 1868 to 1968. See EPA, *First Five Year Report, 2003*, ed. U.S. EPA Region 5 (Chicago: U.S. EPA, 2003), 11.

44. Horace J. Stevens, *The Copper Handbook*, vol. 10 (Houghton, MI: Horace J. Stevens, 1911), 260. Cuprite's chemical formula is Cu_2O.

45. Hyde, *Copper for America*, 20.

46. *Ducktown* is a derivative of a Native American place name for the area.

47. Robert Edward Barclay, *Ducktown Back in Raht's Time* (Chapel Hill: The University of North Carolina Press, 1946), 61.

48. Karen Daniels, *Tennessee's Historic Copper Basin Area: An Overview* (Benton, TN: Polk County Publishing, 1992), 3.

49. Barclay, *Ducktown Back in Raht's Time*, 85.

50. Ibid., 67.

51. Hyde, *Copper for America*, 20–21.

52. Ibid., 20–21; Barclay, *Ducktown Back in Raht's Time*, 149.

53. Many historians believe that William Raht's involvement was in name only and that August Raht actually developed the Bessemer process, derived from a revolu-

tionary procedure Sir Henry Bessemer had developed for exhausting carbon and oxygen from molten iron during the steel-making process.

54. Ken Rush, "Copper Mining Companies of the Basin" (unpublished paper, Ducktown Basin Museum, Ducktown, TN, 1993), 7.

55. "Editor of 1872 Writes About Ducktown Mines," *Cherokee Scout* (Murphy, NC), September 5, 1935, 4. Although this account was likely accurately reported, it is probably an exaggeration because an *Engineering Mining Journal* article about the same process a few years later reports an initial roast charge of only 500 tons or one million pounds (See, Barclay, *Ducktown Back in Raht's Time*, 151).

56. Barclay, *Ducktown Back in Raht's Time*, 151.

57. "Editor of 1872 Writes About Ducktown Mines," 5.

58. "Report of the Secretary of Agriculture for 1875," ed. U.S. Department of Agriculture (Washington, DC: U.S. Department of Agriculture, 1875), 291–94.

59. Robert B. Gordon and Patrick M. Malone, *The Texture of Industry: An Archaeological View of the Industrialization of North America* (New York: Oxford University Press, 1994), 231. This is a loose estimate. Some smelters ran more fuel-intensive reverberatory furnaces; some furnaces were fired directly with wood, which Barclay estimated consumed three times the amount of cordwood as charcoal; and not all smelters ran their furnaces all the time. Assuming that each smelter ran an average of five furnaces each day, then 45,000 bushels of charcoal were consumed per site per month (Barclay, *Ducktown Back in Raht's Time*, 68).

60. One smelter operated in 1854, two in 1855, three smelters in 1856, seven in 1860/61, likely three in 1862, none in 1863, three in 1866–68, two in 1869–71, and one in 1873–1878. Years between were averaged to come up with 63 smelter years.

61. Rush, "Copper Mining Companies of the Basin," 7; Stuart W. Maher, "The Copper-Sulfuric Acid Industry in Tennessee," ed. Division of Geology The State of Tennessee Department of Conservation, Information Circular (Nashville: State of Tennessee, 1966), 4.

62. Christopher J. Schmitz, *World Non-Ferrous Metal Production and Prices 1700–1976* (London: Frank Cass, 1979), 270.

63. "Editor of 1872 Writes About Ducktown Mines," 6.

64. Ibid., 270.

65. Stevens, *The Copper Handbook*, 746 and 1661.

66. Though of German decent, J. B. F. Herreshoff worked in New York and developed several innovative smelting furnaces and processes based on existing designs and principles (Peters, *Modern Copper Smelting*, 266).

67. Robert Edward Barclay, *The Copper Basin, 1890 to 1963* (Knoxville, TN: R. E. Barclay, 1975), 5, and Maher, "The Copper-Sulfuric Acid Industry in Tennessee," 4.

68. Barclay, *The Copper Basin*, 5.

69. Ibid., 33–36.

70. Duncan Maysilles, *Ducktown Smoke: The Fight over One of the South's Greatest Environmental Disasters* (Chapel Hill: University of North Carolina Press, 2011), 68, 123.

71. *A.W. Swain v. Tennessee Copper Co. et al. and Thomas L. Cole v. Ducktown Sulphur, Copper & Iron Co. Limited, et al.* (1903).

72. Daniels, *Tennessee's Historic Copper Basin Area*, 9.

73. Stevens, *The Copper Handbook*, 664 and 747.

74. Barclay, *The Copper Basin*, 6.

75. Stevens, *The Copper Handbook*, 747; Barclay, *The Copper Basin*, 37.

76. Robert Sticht, "Pyritic Smelting—Its History, Principals, Scope, Apparatus, and Practical Results," in *Modern Copper Smelting*, ed. Edward Dyer Peters (New York: The Scientific Publishing, Co., 1898), 400.

77. Hofman, *Metallurgy of Copper*, 144.

78. Maysilles, *Ducktown Smoke*, 119–20.

79. Barclay, *The Copper Basin*, 6–7.

80. Ibid., 13.

81. *State of Georgia v. Tennessee Copper Company and Ducktown Sulphur, Copper, & Iron Company*, 206 U.S. 30 (1907). The other significant event stemming from the 1907 state of Georgia lawsuit was the recognition of environmental damage and the need for some sort of remedy. In the 1920s, reforestation efforts began in the district and continued into the twenty-first century (See, Daniels, *Tennessee's Historic Copper Basin Area*, 13).

82. Barclay, *The Copper Basin*, 39.

83. Ibid., 144–45, and Stevens, *The Copper Handbook*, 1665.

84. Andrew M. Fairlie, *Sulfuric Acid Manufacture* (American Chemical Society's Monograph No. 69), 143–45; Stevens, *The Copper Handbook*, 1666.

85. Fairlie, *Sulfuric Acid Manufacture*, 23–24.

86. Alfred G. Lock and Charles G. Lock, *A Practical Treatise on the Manufacture of Sulphuric Acid* (London: Sampson Low, Marston, Searle, and Rivington, 1879), 100.

87. Fairlie, *Sulfuric Acid Manufacture*, 23.

88. Ibid., 24; Maher, "The Copper-Sulfuric Acid Industry in Tennessee," 18.

89. Barclay, *The Copper Basin*, 8.

90. Stevens, *The Copper Handbook*, 1666.

91. Fairlie, *Sulfuric Acid Manufacture*, 20. Fertilizer production was especially critical in the South, as cotton production consumed considerable soil nutrients.

92. Barclay, *The Copper Basin,* 9.

93. Stevens, *The Copper Handbook*, 1664.

94. Barclay, *Ducktown Back in Raht's Time*, 149; Hofman, *Metallurgy of Copper*, 203; Charles Christensen, "History of Converting," *The Mining World* 33 (1910): 1037.

95. Christensen, "History of Converting," 1036–37.

96. Maher, "The Copper-Sulfuric Acid Industry in Tennessee," 17. Iron sinter is created by first roasting flotation-concentrated iron-rich pyritic ores to release sulfur gas and create calcine (roasted ore concentrate). Gases are transferred to the acid plant, and the calcine is sintered on a traveling grate in a high-heat process that drives off any remaining sulfur and changes the structural characteristic of the material, which assays 70% iron and was sold to the steel industry.

97. Barclay, *The Copper Basin,* 9–11.

98. Ibid., 44.

99. Maher, "The Copper-Sulfuric Acid Industry in Tennessee," 6.

100. Ralph I. Smith, "History of the Early Reduction Plants of Butte, Montana," (Butte: State of Montana Bureau of Mines and Geology: Montana College of Mines, 1953), 1–2.

101. Hyde, *Copper for America*, 81.

102. Smith, "History of the Early Reduction Plants of Butte, Montana," 2–5.

103. Laist, "Seventy-Five Years of Progress in Smelting and Leaching of Ores," 128.

104. Smith, "History of the Early Reduction Plants of Butte, Montana," 5.

105. Ibid., 8; Laist, "Seventy-Five Years of Progress in Smelting and Leaching of Ores," 129.

106. Smith, "History of the Early Reduction Plants of Butte, Montana," 6–8.

107. Hyde, *Copper for America*, 85.

108. Ibid., 85–86; Shovers and others, *Butte & Anaconda Revisited: An Overview of Early-Day Mining and Smelting in Montana*, Special Publication 99 (Butte: Montana Bureau of Mines and Geology; Montana College of Mineral Science and Technology, 1991), 40; Smith, "History of the Early Reduction Plants of Butte, Montana," 6–8.

109. Smith, "History of the Early Reduction Plants of Butte, Montana," 7.

110. Hyde, *Copper for America*, 84.

111. Shovers and others, *Butte & Anaconda Revisited*, 31; Laist, "Seventy-Five Years of Progress in Smelting and Leaching of Ores," 129; Hyde, *Copper for America*, 84.

112. Laist, "Seventy-Five Years of Progress in Smelting and Leaching of Ores," 129; Shovers and others, *Butte & Anaconda Revisited*, 31.

113. Fredric Quivik, "The Anaconda Company Smelters: Great Falls and Anaconda," *The Speculator* 1, no. 2 (1984): 40.

114. Hyde, *Copper for America*, 86.

115. Smith, "History of the Early Reduction Plants of Butte, Montana," 12.

116. Jonathan Leitner, "Red Metal in the Age of Capital: The Political Ecology of Copper in the Nineteenth-Century World-Economy," *Review-Fernand Braudel Center for the Study of Economies, Historical Systems, and Civilizations* 24, no. 3 (2001): 421.

117. Although Anaconda and Great Falls were certainly the largest copper refineries, between 1883 and 1902, ten new refineries opened in the United States, increasing domestic annual capacity from 38,000 tons or 20 percent of U.S. output in 1893 to 347,000 tons, equal to 75 percent of U.S. output in 1906 (See, Hyde, *Copper for America*, 92).

118. Laist, "Seventy-Five Years of Progress in Smelting and Leaching of Ores," 136–37.

119. Ibid., 133.

120. Hofman, *Metallurgy of Copper*, 80 and 87; Laist, "Seventy-Five Years of Progress in Smelting and Leaching of Ores," 133.

121. Smith, "History of the Early Reduction Plants of Butte, Montana," 11–12.

122. Shovers and others, *Butte & Anaconda Revisited*, 21; Hyde, *Copper for America*, 102.

123. Shovers and others, *Butte & Anaconda Revisited*, 35; Hyde, *Copper for America*, 102.

124. Shovers and others, *Butte & Anaconda Revisited*, 35; Hyde, *Copper for America*, 102.

125. Laist, "Seventy-Five Years of Progress in Smelting and Leaching of Ores," 150.

126. Ibid., 135.

127. Anaconda Copper Mining Company, *Copper: From Mine to Finished Product* (New York: Anaconda Copper Mining Company, 1920), 21; Shovers and others, *Butte & Anaconda Revisited*, 39–40.

128. Shovers and others, *Butte & Anaconda Revisited*, 41.

129. Ibid., 41; Quivik, "The Anaconda Company Smelters," 42.

130. Katherine Aiken, "Western Smelters and the Problem of Smelter Smoke," in *Northwest Lands, Northwest People: Readings in Environmental History*, ed. Dale Goble and Paul W. Hirt (Seattle: University of Washington Press, 1999), 505–506.

131. Fredric Quivik, "Smoke and Tailings: An Environmental History of Copper Smelting Technologies in Montana 1880–1930" (PhD diss., University of Pennsylvania, 1998), 399.

132. Ibid., 350.

133. Shovers and others, *Butte & Anaconda Revisited*, 41.

134. Ibid., 41.

135. Ibid., 42–44.

136. Hyde, *Copper for America*, 172.

137. See, Issac Marcosson, *Anaconda* (New York: Dodd, Mead & Company), 1957.

138. Shovers and others, *Butte & Anaconda Revisited*, 13.

139. Ibid., 41.

140. Electric furnaces passed a current through the molten bath of a continuously fed charge, and the natural resistance of the material generated enough heat to maintain high furnace temperatures and continually remelt new charges. Flash smelting blows dried concentrate and oxygen into a hot-furnace hearth and, like pyritic smelting, uses the heat of sulfur and iron oxidation to melt the charge. See, A. K. Biswas and W. G. Davenport, *Extractive Metallurgy of Copper* (New York: Pergamon Press, 1980), 156.

Chapter 3

1. George Grant Hedgcock, "Injury by Smelter Smoke in Southwestern Tennessee." *Journal of the Washington Academy of Sciences* 4, no. 4 (1914): 70–71.

2. Wilton Barnhardt, "The Death of Ducktown." *Discover* 8 (1987): 36.

3. Loraine Lee, Quoted in Barnhardt, "The Death of Ducktown," 41.

4. Gordon Young and Emory Kristoff, "The Tennessee Valley Today," *National Geographic* 143, no. 6 (1973): 858–859. Despite extolling the resurgence of the Tennessee Valley largely through TVA efforts and a caption that described the success of revegetation, the key words to describe the image on the 2008 *National Geographic* website included: desolate landscapes, destruction of woodlands, and nature vs. technology. See, "A Barren Landscape Was Created after Miners Cut Down Forests for Fuel. Location: near Copperhill, Tennessee, Picture Id: 620678, www.nationalgeographicstock.com/ngsimages/explore/, accessed December 10, 2008.

5. M. -L. Quinn, "The Appalachian Mountains' Copper Basin and the Concept of Environmental Susceptibility," *Environmental Management* 15, no. 2 (1991): 183.

6. Stuart W. Maher, "The Copper-Sulfuric Acid Industry in Tennessee," ed. Division of Geology The State of Tennessee Department of Conservation, Information Circular (Nashville, TN: State of Tennessee, 1966), 6–7, 18–19.

7. "Report for the Extent of Lead Contamination at the Former Isabella Lead Chamber Acid Plant, Ducktown, Tennessee," Waggoner Barge, Sumner, & Cannon, Inc. (Nashville, TN: Glenn Springs Holdings, Inc., 2003), 2.

8. Maher, "The Copper-Sulfuric Acid Industry in Tennessee," 6.

9. Tennessee Chemical Corporation will be referenced in the text by TChC so as not to confuse it with the Tennessee Copper Company referred to as TCC.

10. Milton H. Hamilton, "North Potato Creek VOAP Order," in *Site ID 70-508, Case No. 99-0480*, ed. Division of Superfund (Nashville, TN: State of Tennessee, Department of Environment and Conservation, 2001), 5.

11. "Agreement and Covenant Not to Sue (Boliden Intertrade)," ed. U.S. EPA (Washington, D.C.: U.S. EPA, 1990), 1; John H. Hankinson Jr., "Administrative Order on Consent for Removal Action at the North Potato Creek Watershed," in *CERCLA Docket No. 01-11-C*, ed. Region 4 U.S. EPA (Atlanta, GA: U.S. EPA, 2001), 4; "Copper Basin Mining District," U.S. EPA, accessed March 9, 2009. http://www.epa.gov/region4/waste/copper/

12. "Report of Inventory and Characterization of Former Mining and Processing Areas and Materials Located on Public and Private Lands," ed. Waggoner Barge, Sumner, & Cannon, Inc. (Ducktown, TN: Glenn Springs Holdings, Inc., 2004), 2–2.

13. "Community Involvement Plan for the Copper Basin Mining District," ed. Community Involvement Partnership (Polk County, TN: U.S. EPA, Tennessee Department for Environment and Conservation, Glenn Springs Holdings, Inc., 2001), 3–4.

14. "A Future Use Plan for Redevelopment of the Abandoned Mine Lands in Copperhill Tennessee," ed. Waggoner Barge, Sumner, and Cannon, Inc. (Ducktown, TN: Glenn Springs Holdings, Inc., 2003), 3–4.

15. Southland Company is the parent to the 7/11 chain of convenience stores and sought ownership of Cities Service to provide a steady supply of gasoline to its stores. "Citgo Company History," CITGO Petroleum Company, accessed March 4, 2009. www.citgo.com/aboutcitgo/companyhistory.jpg

16. "Copper Basin," accessed March 7, 2009, http://www.glennsprings-copperbasinproject.com/history.htm

17. Bode J. Morin, personal communication to Ducktown mayor, 22 October 2008; Bode J. Morin, personal communication to a federal official close to the cleanup, 23 October 2008; "Copper Basin Mining District Case Study: Use of Cooperative Agreements Toward a Common Goal," ed. U.S. EPA (Washington, DC: U.S. EPA, 2005), 6.

18. Karen Daniels, *Tennessee's Historic Copper Basin Area: An Overview* (Benton, TN: Polk County Publishing, 1992), 13. Twentieth-century environmental awareness in Montana, however, was hard to gauge because, by 1930, the Anaconda Company held

considerable political power in the state, owned several newspapers, and shared board members with other regional corporations. Although similar power structures existed with the largest smelters in Swansea, neither Michigan nor Tennessee mining/smelting companies enjoyed such influence.

19. B. C. Hall and C.T. Wood, *The South* (New York: Scribner, 1995), 224.

20. This quick study was conducted by searching for images and postcards at the Library of Congress website, accessed 3 April 2009, www.memory.loc.gov and www.google.com with under the terms *Ducktown, Tennessee; Copperhill, Tennessee; Butte, Montana; Anaconda, Montana; Houghton, Michigan; and Calumet, Michigan.*

21. Quinn, "The Appalachian Mountains' Copper Basin and the Concept of Environmental Susceptibility," 185–86.

22. Ibid., 189.

23. Jack A. Muncy, "A Plan for Revegetation Completion of Tennessee's Copper Basin," ed. Division of Land and Economic Resources (Norris, TNe: Tennessee Valley Authority, 1986), 6.

24. Edward A. Johnson, "Industrial Destruction Reversed at Copper Basin." *National Woodlands* 25, no. 2 (2002):10–13.

25. "Engineering Report on Land Reclamation: Copper Basin- Polk County, Tennessee," Cities Service Company (Copperhill, TN: Cities Service Company, 1973), 7.

26. Muncy, "A Plan for Revegetation Completion of Tennessee's Copper Basin," 9.

27. Ibid., 10.

28. M. -L. Quinn, "Should All Degraded Landscapes Be Restored? A Look at the Appalachian Copper Basin," *Land Degradation & Rehabilitation* 3 (1992): 118.

29. Muncy, "A Plan for Revegetation Completion of Tennessee's Copper Basin," 8.

30. Ibid., 12.

31. Kenneth J. Seigworth, "Ducktown—a Postwar Challenge: Can Man Correct a Century of Land Abuse in Tennessee's Copper Basin? A Start Has Been Made," *American Forests* 49 (November 1943): 522; Muncy, "A Plan for Revegetation Completion of Tennessee's Copper Basin," 19; "Copper Basin Reclamation Project," Tennessee Valley Authority, accessed March 11, 2009, http://www.tva.gov/environment/land/copper.htm

32. Mouzon Peters, "Healing the Scars in the Copper Basin," *Chattanooga Sunday Times*, September 2, 1956, 1.

33. Muncy, "A Plan for Revegetation Completion of Tennessee's Copper Basin," 12.

34. Ibid., 13.

35. Johnson, "Industrial Destruction Reversed at Copper Basin," 29.

36. Bode J. Morin, personal communication to museum director, 24 October 2008.

37. Bode J. Morin, personal observation, 24 October 2008.

38. "Browning Is Speaker Saturday at Conservation League Meet," *Copper City Advance*, May 25, 1950, 1.

39. "'Must Conserve Resources If We Survive,' Browning," *Copper City Advance*, June 1, 1950, sec. 1.

40. Peters, "Healing the Scars in the Copper Basin," 1.

41. See Seigworth, "Ducktown—a Postwar Challenge: Can Man Correct a Century of Land Abuse in Tennessee's Copper Basin? A Start Has Been Made"; Peters, "Healing the Scars in the Copper Basin"; Clay Grady, "Copper Basin Cover-Up," *Landscape Architecture* 73, no. 4 (1983): 49–55; Young and Kristoff, "The Tennessee Valley Today"; Wilton Barnhardt, "The Death of Ducktown," *Discover* 8 (1987): 35–42; Johnson, "Industrial Destruction Reversed at Copper Basin"; Chris Dixon, "Up from the Mines in Tennessee," *New York Times*, December 14, 2007.

42. Quinn, "Should All Degraded Landscapes Be Restored? A Look at the Appalachian Copper Basin," 130–31.

43. See, Donald Hardesty, "Issues in Preserving Toxic Wastes as Heritage Sites," *The Public Historian* 23, no. 2 (2001): 19.

44. See David Robertson, "Heaps of History: Toluca and the Historic Longwall Mining District," *Journal of Illinois History* 3 (Autumn 2000): 184.

45. See, Frederic L. Quivik, "Integrating the Preservation of Cultural Resources with Remediation of Hazardous Materials: An Assessment of Superfund's' Record," *The Public Historian* 23, no. 2 (2001): 47.

46. Quoted in Dixon, "Up from the Mines in Tennessee."

47. "Community Involvement Plan for the Copper Basin Mining District," 4–5.

48. Bode J. Morin, personal communication to a person close to the cleanup who desired anonymity, October 22, 2008.

49. Quoted in Dixon, "Up from the Mines in Tennessee."

50. Bode J. Morin, personal communication to Ducktown mayor, 22 October 2008.

51. "Tennessee Historical Commission State Owned Sites," Tennessee Department of Environment and Conservation, accessed March 9, 2009, www.state.tn.us/environmental/hist/stateown/ducktown.shtml

52. Ron Clayton, "Ducktown Basin Museum Hits 30 with Plans to Keep Growing," *Chattanooga Times Free Press*, October 13, 2008; Bode J. Morin, personal communication to museum director, 24 October 2008.

53. See, Karen Daniels, "National Register of Historic Places Registration Form: Central Headframe," United States Department of the Interior (Washington, DC: Na-

tional Park Service, 1992); Karen Daniels, "National Register of Historic Places Registration Form: Copperhill Historic District," United States Department of the Interior (Washington, DC: National Park Service, 1992); Karen Daniels, "National Register of Historic Places Registration Form: Ducktown Historic District," United States Department of the Interior (Washington, DC: National Park Service, 1992); Karen Daniels, "National Register of Historic Places, Multiple Property Documentation Form: Historic Resources of the Tennessee Copper Basin," United States Department of the Interior (Washington, DC: National Park Service, 1992).

54. The Copperhill plant was still operational at the time of the memorandum of understanding (MOU) and thus fell under other jurisdictional legislation aimed more at controlling active waste streams than remediating past pollution.

55. "Report of Inventory and Characterization of Former Mining and Processing Areas and Materials Located on Public and Private Lands," 4–13.

56. "Copper Basin Mining District Case Study: Use of Cooperative Agreements toward a Common Goal," 3.

57. "Copper Basin Mining District," U.S. EPA, accessed March 13, 2009, http://www.epa.gov/region4/waste/copper/

58. "Copper Basin," http://www.glennsprings-copperbasinproject.com/history.htm, accessed March 7, 2009.

59. See, Timothy Fields, "Memorandum of Understanding between Tennessee Department of Environment and Conservation, U.S. EPA and Oxy, Inc.," in *CECRLA Docket No. 01-10-C*, ed. U.S. EPA Region 4 (Atlanta, GA: U.S. EPA Region 4, 2001).

60. See ibid.

61. "Copper Basin Mining District Case Study: Use of Cooperative Agreements toward a Common Goal," 6; Bode J. Morin, personal communication to Ducktown mayor, 22 October 2008.

62. "Copper Basin Mining District," U.S. EPA, accessed March 13, 2009, http://www.epa.gov/region4/waste/copper/

63. Bode J. Morin, personal communication to museum director, 24 October 2008.

64. "Copper Basin Mining District," U.S. EPA, accessed March 13, 2009, http://www.epa.gov/region4/waste/copper/

65. "Community Involvement Plan for the Copper Basin Mining District," coversheet.

66. Ibid., 4–4.

67. Ibid., 4–5.

68. "A Future Use Plan for Redevelopment of the Abandoned Mine Lands in Copperhill Tennessee," 1.

69. Ibid., 10–12.

70. Ibid., 12.

71. Ibid., 12.

72. "USA State & County Quickfacts," U.S. Census Bureau, March 24, 2009, http://quickfacts.census.gov

73. Bode J. Morin, personal communication to museum director, 24 October 2008.

74. Bode J. Morin, personal communication to Ducktown mayor, 22 October 2008.

75. "Report for the Extent of Lead Contamination at the Former Isabella Lead Chamber Acid Plant, Ducktown, Tennessee," 2.

76. "A Future Use Plan for Redevelopment of the Abandoned Mine Lands in Copperhill Tennessee," detached one-page summary.

77. Ibid., detached one-page summary.

78. Bode J. Morin, personal communication to Ducktown mayor, 22 October 2008.

Chapter 4

1. Dashiell Hammett, *Red Harvest* (New York: Vintage Books, 1972), 3–4.

2. Tom Dickson, quoted by Laurie Mercier in *Anaconda: Labor, Community, and Culture in Montana's Smelter City* (Urbana: University of Illinois Press, 2001), 216.

3. Keith Haugland, "Stubborn Stack Dies Slowly," *Great Falls Tribune*, September 19, 1982.

4. Bode J. Morin, personal communication between author and fellow researcher, 12 November 2009.

5. Gallows frames is the colloquial term for the structures at the opening of an underground mine, called head frames in many other parts of the country.

6. The city of Butte was settled and developed on Butte Hill, intermixed with the mines. The central business district, commonly referred to as "downtown" in many cites, is called "uptown" in Butte.

7. "Welcome to Anaconda," Community of Anaconda, Montana, accessed April 2, 2009, http://www.anacondamt.org/; Bode J. Morin, personal observation, November 2008.

8. A life-size statue of Marcus Daly sculpted by Augustus Saint-Gaudens in 1906 sat in a prominent uptown intersection until a traffic accident forced the removal of the sculpture to the Montana Tech campus. It is preserved and maintained and, as such, has heritage value, but maybe not as much as it did when it was located in a more prominent location.

9. See, Maney Telefilm, *Old Works Vision to Reality*, VHS (Butte, MT: ARCO Environmental Remediation) (ca. 1998).

10. William Wyckoff, "Postindustrial Butte," *Geographical Review* 85, no. 4 (1996): 45.

11. Extrapolated from Janet L. Finn, *Tracing the Veins: Of Copper, Culture, and Community from Butte to Chuquicamata* (Berkeley: University of California Press, 1998), 247; Otis E. Young, "The American Copper Frontier, 1640–1893," *The Speculator* 1, no. 2 (1984): 7.

12. Isaac Frederick Marcosson, *Copper Heritage: The Story of Revere Copper and Brass Incorporated* (New York: Dodd Mead, 1955), 143. See also, Marcosson, *Anaconda* 136–93.

13. Fredric L. Quivik, "The Anaconda Company Smelters: Great Falls and Anaconda," *The Speculator* 1, no. 2 (1984): 44.

14. Ibid., 44.

15. Mary Elisabeth Curran, "The Contested Terrain of Butte, Montana: Social Landscapes of Risk and Resiliency" (master's thesis, University of Montana, 1996), 93. ACM avoided trial from the lawsuits brought by the United States by negotiating some changes to its operations and exchanging good land for damaged.

16. David M. Emmons, "'The Price of 'Freedom': Montana in the Late and Post-Anaconda Era," *Montana: The Magazine of Western History* 44, no. 4 (1994): 68.

17. Quivik, "The Anaconda Company Smelters: Great Falls and Anaconda," 44; Michael P. Malone, Richard B. Roeder, and William L. Lang, *Montana: A History of Two Centuries*, Rev. ed. (Seattle: University of Washington Press, 1991), 247; Wyckoff, "Postindustrial Butte," 478.

18. Malone, Roeder, and Lang, *Montana: A History of Two Centuries*, 249.

19. Emmons, "'The Price of 'Freedom,'" 68.

20. Finn, *Tracing the Veins*, 247.

21. The ACM block caving project was announced in 1947, yielded its first copper in 1952, and ended in 1961, although other underground mining continued through the late 1960s. See, Marcosson, *Copper Heritage; the Story of Revere Copper and Brass Incorporated*, 274; Brian Shovers and others, *Butte & Anaconda Revisited: An Overview of Early-Day Mining and Smelting in Montana*, 13 and 26; and Malone, Roeder, and Lang, *Montana : A History of Two Centuries*, 249.

22. *Mining the Berkeley Pit 1955–1982* (Butte, MT: Butte Chamber of Commerce, 2004; Finn, *Tracing the Veins*, 191.

23. Open-pit mining in Butte continued beyond 2008 buoyed by high, prerecession Asian demand, albeit on a much smaller scale and with efficiencies requiring far fewer people.

24. Bode J. Morin, personal observation, November 2008.

25. "County Population Census Counts 1900–90," U.S. Census Bureau, accessed April 4, 2009, http://www.census.gov/population/www/censusdata/cencounts/index.html

26. Emmons, "'The Price of 'Freedom,'" 69; Malone, Roeder, and Lang, *Montana: A History of Two Centuries*, 250.

27. Malone, Roeder, and Lang, *Montana: A History of Two Centuries*, 250.

28. Emmons, "'The Price of 'Freedom,'" 69.

29. Eugene C. Tidball, "What Ever Happened to the Anaconda Company?" *Montana: The Magazine of Western History* 47, no. 2 (1997): 68.

30. Shovers and others, *Butte & Anaconda Revisited: An Overview of Early-Day Mining and Smelting in Montana*, 61.

31. "Anaconda Copper Smelter May Switch to Coal," *Chemical Week* 123, (1980): 61.

32. Vera Haffey, "Anacondans Remember Day the Smelter Closed," *Montana Standard*, September 29, 2005, A1 and A6.

33. "EPA Chief Says Closure a Surprise," *The Montana Standard*, October 1, 1980.

34. "Role of Clean Air Act Requirements in Anaconda Copper Company's Closure of Its Montana Smelter and Refinery," ed. Region 8 U.S. EPA (Denver, CO: U.S. EPA, 1981), 5–7.

35. Tidball, "What Ever Happened to the Anaconda Company?" 63.

36. Ibid., 67.

37. Shari Meets, "Anaconda Co. Closes Smelter," *Montana Standard*, September 30, 1980, 1.

38. "Role of Clean Air Act Requirements in Anaconda Copper Company's Closure of Its Montana Smelter and Refinery," U.S. EPA, 1981, 8.

39. Ibid., 13.

40. Bob McCarthy, "Arco's Lethal Legacy," *Montana Eagle* 1981, 4.

41. Hugh van Sweringen, "Shutdown Price Tag Will Be $40 Million Annually," *The Montana Standard*, September 30, 1980, 1.

42. Lynda Chavez, "When Arco Left Town," *New York Times*, July 25, 1982, 15.

43. Carrol Pursell, "The Rise and Fall of the Appropriate Technology Movement in the United States, 1965–1985," *Technology and Culture* 34, no. 3 (1993): 630.

44. The town of Mill Creek sat just a few miles from the Washoe smelter stack directly under the smoke plume. In the mid 1980s, the EPA discovered very high levels of arsenic inside homes and local soils, and the Center for Disease Control tested the town's children, who showed significantly elevated levels of arsenic in their blood. Although the first recourse was intended to be a temporary displacement while the EPA

supervised remediation, the government agencies soon discovered levels too high, and ultimately completely leveled the town.

45. Michael P. Malone, "The Close of the Copper Century," *Montana: The Magazine of Western History* 35, no. 2 (1985): 69.

46. Emmons, "'The Price of 'Freedom,'" 71.

47. See, Duncan Adams, "Did Toxic Stew Cook the Goose?," *High Country News*, December 11, 1995, accessed August 10, 2009, www.hcn.org/issues/49/1520

48. Cited in Florence Williams, "Butte, Montana, Seeks a New Life," *High Country News*, November 29, 1993, 2.

49. Peter Greenberg, *Don't Go There!: The Travel Detective's Essential Guide to the Must-Miss Places of the World* (Emmaus, PA: Rodale, 2009), 33.

50. David M. Emmons, "United States District Court, District of Montana, Butte Division; United States of America v. Atlantic Richfield Company, et al; Expert Report of David Emmons, Ph.D." (Butte, MT: 1997), 2.

51. Fred Quivik, "Smoke and Tailings," 213.

52. Marcosson, *Copper Heritage: The Story of Revere Copper and Brass Incorporated*, 105.

53. Gordon Morris Bakken, "Montana, Anaconda, and the Price of Pollution," *The Historian* 69, no. 1 (2007): 42–45.

54. Emmons, "United States District Court, District of Montana, Butte Division; United States of America v. Atlantic Richfield Company, et al; Expert Report of David Emmons, Ph.D.," 8.

55. Quivik, "Smoke and Tailings," 457–459.

56. Ibid., 460 and 489.

57. Dan Baum and Margaret L. Knox, "We Want People Who Have a Problem with Mine Wastes to Think of Butte," *Smithsonian* 23, no. 8 (1992): 48.

58. "History of the Museum," World Museum of Mining, accessed April 2, 2009, www.miningmuseum.org/historyofwmm.html, , and personal observation November 2008.

59. Personal Communication with Butte-Silver Bow Archivist, November 11, 2008.

60. Shovers et al., *Butte & Anaconda Revisited: An Overview of Early-Day Mining and Smelting in Montana*, 1.

61. See: Donald F. Dosch, "National Register of Historic Places Inventory - Nomination Form for Butte Historic District," ed. United States Department of the Interior (Washington, D.C.: National Park Service, 1972).

62. Personal Communication with Butte-Silver Bow archivist, November 11, 2008.

63. Brian Shovers, "Remaking the Wide-Open Town: Butte at the End of the Twentieth Century," *Montana: The Magazine of Western History* 48, no. 3 (1998), 44.

64. Bode J. Morin, personal communication to local residents, 13 November 2008.

65. Shovers, "Remaking the Wide-Open Town," 44.

66. Wyckoff, "Postindustrial Butte," 489.

67. Shovers, "Remaking the Wide-Open Town," 46–47.

68. Ibid., 47.

69. Dale Martin and Brian Shovers, "Butte, Montana: An Architectural and Historical Inventory of the National Landmark District" (Butte, MT: Butte Historical Society, 1986), 1.

70. "Butte, Montana: A Project Report," ed. Heritage Conservation and Recreation Service (Butte, Montana National Park Service, 1980), 7. The HCRS was abolished in 1981and HAER was transferred to the NPS.

71. Bode J. Morin, personal communication to a state library manager, 17 November 2008.

72. "Minutes of the Meeting, May 26, 2005," ed. Butte-Silver Bow Local Government Study Commission (Butte, MT: 2005), 9. The URA loaned money for building acquisitions, redevelopment, and other improvements to owners of neglected buildings and houses in the historic district to fund improvements that ultimately raised the value of the building. URA also made facade improvement grants and invested in infrastructure improvements.

73. Bode J. Morin, personal communication to Butte-Silver Bow planner, 20 October 2008.

74. Pursell, "The Rise and Fall of the Appropriate Technology Movement in the United States, 1965–1985," 630.

75. Ibid., 636.

76. Bode J. Morin, personal communication to Butte-Silver Bow archives staff, 11 November 2008, and personal communication to state library manager, 17 November 2008.

77. Bode J. Morin, personal communication to state library manager, 11 November 2008, and personal communication to Butte-Silver Bow archivist, 11 November 2008.

78. Shovers, "Remaking the Wide-Open Town," 47.

79. Bode J. Morin, personal communication to Anaconda-Deer Lodge planner, 17 November 2008.

80. See, *Old Works Vision to Reality*

81. Jerrold J. Marcus, "Butte: Richest Hill on Earth and Costliest Mine Superfund Site," *Engineering and Mining Journal* 201, no. 2 (2000): 33. The 30 billion pounds of metals does not include MRI that produced 1.34 billion pounds through 1999.

82. "Superfund Program: Silver Bow Creek/Butte Area," U.S. EPA, accessed April 18, 2009, http://www.epa.gov/region8/superfund/mt/sbcbutte/index.html

83. "Superfund Program: Anaconda Co. Smelter Site," U.S. EPA, accessed April 18, 2009, http://www.epa.gov/region8/superfund/mt/anaconda/index.html84. Remediation of a fourth regional Superfund site, the Montana Pole and Treating Plant Site was also funded by ARCO although its contamination came primarily from the treatment of wood poles for bridge and road construction, and utility needs from the 1940s–1980s not directly from mining activities.

85. Bode J. Morin, personal communication to an EPA official, 20 November 2008.

86. Emmons, "United States District Court, District of Montana, Butte Division; United States of America v. Atlantic Richfield Company, et al; Expert Report of David Emmons, Ph.D.," 96; "State of Montana v. Atlantic Richfield Company, Consent Decree," ed. Helena Division United States District Court for the District of Montana (Helena, MT: United States District Court, 1998).

87. Emmons, "United States District Court, District of Montana, Butte Division; United States of America v. Atlantic Richfield Company, et al; Expert Report of David Emmons, Ph.D.," 53.

88. United States District Court for Montana, "United States v. Atlantic Richfield Company, State of Montana v. Atlantic Richfield Company, Streamside Tailings Operable Unit and Federal and Tribal Natural Resource Damages Consent Decree," (Helena, MT.: United States District Court for Montana, 1998), 2; "Summary of 2008 Settlement of Clark Fork River Remediation and Natural Resources Damages Claims and Related Restoration Plans," ed. Montana Department of Justice (Helena, MT: Natural Resource Damage Program, 2008), 1–3.

89. "Upper Clark Fork River Basin Restoration Plan Procedures and Criteria," ed. Natural Resource Damage Program State of Montana (Helena, MT: State of Montana, 2000), 7.

90. "Summary of 2008 Settlement of Clark Fork River Remediation and Natural Resources Damages Claims and Related Restoration Plans," 1.

91. Sherry Devlin, "Arco Recommends Leaving Milltown Sediment," *Missoulian*, November 6, 2001.

92. Karen Dorn Steele, "Superfund Revived Butte," *Spokesman Review*, July 28, 2002.

93. Bode J. Morin, personal communication to an EPA official, 11 November 2008.

94. Marcus, "Butte: Richest Hill on Earth and Costliest Mine Superfund Site," 41–42.

95. Curran, "The Contested Terrain of Butte, Montana: Social Landscapes of Risk and Resiliency," 269.

96. Jeffery St. Clair, "Something About Butte," in *Red State Rebels: Tales of Grassroots Resistance in the Heartland*, ed. Joshua Frank and Jeffery St. Clair (Oakland, CA: AK Press, 2008), 93.

97. Fredric L. Quivik, "Integrating the Preservation of Cultural Resources with Remediation of Hazardous Materials: An Assessment of Superfund's Record," *The Public Historian* 23, no. 2 (2001): 48, 51.

98. Bode J. Morin, personal communication to Anaconda planner, 17 November 2008.

99. Bode J. Morin, personal communication to Butte-Silver Bow Planner, 20 November 2008.

100. "Council Bill No. 238 ("1985 Preservation Ordinance")," in *Ordinance 238*, ed. Butte-Silver Bow Montana (Butte, MT: 1985), 8.

101. Ibid., 7.

102. Martin and Shovers, "Butte, Montana: An Architectural and Historical Inventory of the National Landmark District," 1.

103. See, Ibid.; Mark Fiege, Fredric L. Quivik, and Brian Shovers, "Industrial Heritage of Butte and Anaconda: An Analysis of the Historical Significance of the Surviving Physical Features of the Anaconda Copper Mining Company," (Butte, MT: Butte Historical Society, 1985).

104. "The Butte-Anaconda Historical Park System Master Plan," ed. Renewable Technologies Inc. (Butte, MT: Butte Historical Society, 1985), i.

105. Bode J. Morin, personal communication to Butte-Silver Bow archivist, 11 November 2008.

106. "The Butte-Anaconda Historical Park System Master Plan," 56.

107. "Programmatic Agreement among the Environmental Protection Agency Region VIII Montana Office, the Advisory Council on Historic Preservation, the Montana State Historic Preservation Office, Montana Department of Health and Environmental Sciences, Local Governments of Butte/Silver Bow and Anaconda/Deer Lodge, Walkerville, and Arco Regarding Development of and Implementation of the Upper Clark Fork

River Basin Regional Historic Preservation Plan," ed. Environmental Protection Agency (Butte, MT: 1992), 1.

108. See, "Regional Historic Preservation Plan: Butte-Anaconda Corridor," (Butte, MT: Regional Historic Preservation Plan Joint Committee, 1993).

109. See, Dosch, "National Register of Historic Places Inventory - Nomination Form for Butte Historic District," and "National Register of Historic Places Registration Form: Butte-Anaconda Historic District-National Historic Landmark Nomination," ed. Department of the Interior (Butte, MT: National Park Service, 2005).

110. Butte Citizens for Preservation and Restoration (CPR) and the Butte Restoration Alliance are both private nonprofit member organizations. CPR provides grants, education, workshops, and coordination with other public and private groups to promote preservation in the National Historic Landmark District (Accessed May 3, 2009, www.buttecpr.org). The Restoration Alliance provides expert advice on spending and development to promote safety, neighborhood enhancement, historical and cultural preservation, and recreation opportunities (Accessed May 3, 2009, www.butterestorationalliance.org). Certified Local Governments are local governments certified by the NPS for a commitment to preservation and are thus eligible for certain preservation grants and funding.

111. See, Baum and Knox (1992), Dobbs (1996), Baum (1997), Robbins (1998), Marcus (2000), Lurgio (2001), Robbins (2005), St. Clair (2008), Egan (1997), and Robbins (1997).

112. Fredric L. Quivik, "Landscapes as Industrial Artifacts: Lessons from Environmental History," *IA: The Journal of the Society for Industrial Archeology* 26, no. 2 (2000): 61.

113. *A Story of Rebirth* (Butte, MT: ARCO, c. 2004).

114. Bode J. Morin, personal communication to an EPA official, 11 November 2008.

115. "The Butte-Anaconda Historical Park System Master Plan," 74.

116. Ibid., 82 and 153.

117. "Regional Historic Preservation Plan: Butte-Anaconda Corridor," V-21.

118. See, Quivik, "Landscapes as Industrial Artifacts."

119. *A Story of Rebirth.*

120. "Old Works Engineering Evaluation/Cost Analysis-Preliminary Site Characterization," ed. PTI Environmental (Bellevue, WA: ARCO, 1991), 20; "Old Works/East Anaconda Smelter Case Study," accessed September 20, 2008, www.epa.gov/superfund/programs/recycle_old/success/casestud/anaccsi.htm

121. "Old Works Operable Unit: Engineering Evaluation/Cost Analysis Work Plan," ed. U.S. Bureau of Reclamation (Billings, MT: U.S. Environmental Protection Agency, 1988), 18–22.

122. "The Butte-Anaconda Historical Park System Master Plan," 68, 82, and 131.

123. *Old Works Vision to Reality*. Video. Sand + slag = sland.

124. "Old Works/East Anaconda Smelter Case Study," accessed September 20, 2008, www.epa.gov/superfund/programs/recycle_old/ success/casestud/anaccsi.htm

125. "Final Work Plan for Anaconda Smelter RI/FS," ed. CH2M Hill (Anaconda, Montana: U.S. EPA, 1988), 1–2.

126. Ibid., 1–4.

127. Ibid., 1–5.

128. "Superfund Program, Anaconda Co. Smelter Site," accessed July 8, 2009, www.epa.gov/unix0008/superfund/mt/anaconda/index.html

129. "General Work Plan, Anaconda Smelter NPL Site, Attachment I," (Anaconda, MT: U.S. EPA Region 8, 1988), 28–32.

130. Ibid., 30 and 32.

131. Bode J. Morin, personal communication to leaders of the Preserve the Stack Committee, 20 November 2008.

132. Bode J. Morin, personal observation, November 2008.

133. "The Butte-Anaconda Historical Park System Master Plan," 68.

134. Bode J. Morin, personal conversations with Michigan Technological University economists, 2009.

Chapter 5

1. C&H ended mining in 1969 and, although Quincy ended native copper mining in 1945, its mills continued through 1967 and its smelter until 1971. Copper Range ended native copper mining in 1967, but its copper sulfide operation continued in Ontonagon County until 1997.

2. Charles K. Hyde, *Copper for America*, 66.

3. Ibid., 60 and 66. Perception of Michigan copper changed, too, in the late nineteenth century. Once considered the standard metal for electrical use because of its natural purity, Michigan copper ultimately became just another electrical-grade metal because the purity of electrolytic-refined copper easily matched the 99.9% pure Michigan metal. In fact in 1892, a New York electrolytic-refining copper firm began marketing "Nichols-Lake Substitute" copper. Curtis Cravens, *Copper on the Creek: Reclaiming an Industrial History* (New York: Place in History, 2000), 11.

4. Larry Lankton, *Cradle to Grave,* 245.

5. Hyde, *Copper for America,* 76–79.

6. Lankton, *Cradle to Grave,* 244.

7. Ibid., 244–45.

8. Ibid., 249–50. C&H continued to process its tailings until WWII.

9. Ibid., 259.

10. Ibid., 262.

11. The site still stood in 2009 as a large and open ruin with no complete structures except for the original office building that now serves as a residence.

12. Dan Trepal, "The Calumet & Hecla Smelting Works: 1887–2006," 24.

13. Bode J. Morin, personal observations, 2006–9.

14. Lankton, *Beyond the Boundaries,* 12.

15. Although lawsuits against Michigan Copper Mines for property damages were rare if they occurred at all, late nineteenth-century tailings deposits into Portage Lake ultimately resulted in significant obstructions to navigation requiring dredging by the Army Corps of Engineers to maintain the waterway. Strong (2002) demonstrated that significant tailings deposits from 1882 to 1886 had reduced the already-constricted navigable water by more than 66% at one location in east Houghton. See, Katherine Strong, "Shoreline Change Along the Portage Canal" (Unpublished graduate paper available at Michigan Technological University Archives) (Houghton, Mich.: Michigan Technological University, 2002), 5.

16. Aaron Shapiro, "Air Conditioned by the Cool Breezes of Lake Superior: Vacationing in Michigan's Copper Country," in *New Perspectives on Michigan's Copper Country,* ed. Alison K. Hoagland, Erik C. Nordberg, and Terry S. Reynolds (Hancock, MI: Quincy Mine Hoist Association, 2007), 135.

17. Ibid., 137.

18. Ibid., 144–45.

19. *Cloverland Tourist Guide* (Menomonee, MI: The Herald-Leader Company, 1926), 32.

20. "Arcadian Mining Property Once the Rival of C&H," *Daily Mining Gazette,* June 24, 1978.

21. unknown, "Copperama" (ca. 1960) (Available at the Michigan Technological University Archives).

22. "Options for National Park Service Involvement in the Management of Historic Copper Mining Resources on Michigan's Keweenaw Peninsula," ed. Midwest Region National Park Service (Omaha, NE: National Park Service, 1988), 3.

23. Erik C. Nordberg, "Quincy Mine Hoist Association Anniversary Banquet," booklet available at Michigan Technological University Archives (Hancock, MI: Quincy Mine Hoist Association, 2008).

24. "Options for National Park Service Involvement in the Management of Historic Copper Mining Resources on Michigan's Keweenaw Peninsula," 3–4.

25. Ibid., 4.

26. "Final General Management Plan and Environmental Impact Statement," ed. Keweenaw National Historical Park Michigan (Calumet, MI: U.S. Department of the Interior, 1998), 7.

27. "Options for National Park Service Involvement in the Management of Historic Copper Mining Resources on Michigan's Keweenaw Peninsula," 6.

28. "Final General Management Plan and Environmental Impact Statement," 50.

29. Ibid., 13–14.

30. Ibid., 175.

31. "Keweenaw Heritage Sites," National Park Service, accessed June 25, 2009. http://www.nps.gov/kewe/parkmgmt/keweenaw-heritage-sites.htm

32. James M. Olson, *Michigan Environmental Law: A Guide to Environmental Law* (Traverse City, MI: Neahtawanta Press, 1981), 10.

33. "Second Five-Year Review Report for the Torch Lake Superfund Site," ed. Region 5 U.S. EPA (Chicago, IL: U.S. EPA, 2008), 12 and 18.

34. Ibid., 17.

35. "First Five-Year Review Report for the Torch Lake Superfund Site," ed. Region 5 U.S. EPA (Chicago, IL: U.S. EPA, 2003), 13.

36. "EPA Completes Construction at Torch Lake," ed. Region 5 U.S. EPA (Chicago, IL: U.S. EPA, 2006), 1.

37. Ibid., 15.

38. "Proposed Plan for Record of Decision Amendment for Operable Unit 3 at Torch Lake Superfund Site," ed. Region 5 U.S. EPA (Chicago, IL: U.S. EPA, 2008), 2.

39. Ibid., 3.

40. "First Five-Year Review Report for the Torch Lake Superfund Site," 16–18.

41. "Responsiveness Summary, Torch Lake Superfund Site, Operable Units I and II, Houghton County, Michigan," ed. Region 5 U.S. EPA (Chicago, IL: U.S. EPA, 1992), RS-30.

42. "Proposed Plan for Record of Decision Amendment for Operable Unit 3 at Torch Lake Superfund Site," 3.

43. "Declaration for the Record of Decision, Torch Lake Site, Michigan, Operable Units I and II," ed. Region 5 U.S. EPA (Chicago, IL: U.S. EPA, 1992), 19, and "EPA Proposes

Cleanup Plan for Quincy Smelter Erosion," ed. Region 5 U.S. EPA (Chicago, IL: U.S. EPA, 2008), 1.

44. Kurt Hauglie, "EPA to Host Smelter Hearing," *Daily Mining Gazette*, January 12, 2009.

45. James Corless, "The Significance and Status of the Quincy Smelting Works, Franklin Township, Michigan (White Paper)," ed. Keweenaw National Historical Park (unpublished paper, Keweenaw National Historical Park, 2008), 1.

46. "Programmatic Agreement among the Michigan State Historical Preservation Officer, Michigan Technological University and Keweenaw National Historical Park, National Park Service (U.S. Department of the Interior), Regarding the A.E. Seaman Mineralogical Museum Adaptive Reuse Project, Er06-107," ed. Michigan State Historic Preservation Office (Lansing, MI: MISHPO, 2006), 2; Corless, "The Significance and Status of the Quincy Smelting Works, Franklin Township, Michigan (White Paper)," 7.

While the existence of the Programmatic Agreement does not implicate the Michigan SHPO in avoiding Section 106 of NHPA responsibility for the Quincy Smelter site, it demonstrates a willingness to enter into agreements to allow other interested parties, such as the NPS, to enforce compliance at sites quite distant from the state capital. The White Paper written by NPS officials suggested that NPS "exercised its authority granted by the National Historic Preservation Act to comment on and influence the activities of a federally funded project with potential adverse affects on a national historic landmark," (p. 6–7). These responsibilities would be the same for the SHPO under those circumstances.

47. "Record of Decision, Decision Summary, Torch Lake Sites, Operable Units I & II, Houghton County Michigan," ed. Region 5 U.S. EPA (Chicago, IL: U.S. EPA, 1992), 27.

48. Ibid., 27.

49. "Final Remedial Investigation Report, Operable Unit III," ed. Region 5 U.S. EPA (Chicago, IL: Donohue and Associates, 1992), 4–3.

50. "Record of Decision, Decision Summary, Torch Lake Sites, Operable Units I & II, Houghton County Michigan," 27.

51. "Declaration for the Record of Decision, Torch Lake Site, Michigan, Operable Units I and II," 3.

52. Corless, "The Significance and Status of the Quincy Smelting Works, Franklin Township, Michigan (White Paper)," 7.

53. Ibid., 7.

54. "Letter from EPA Region 5 to Frank Fiala, National Park Service-U.S. Department of the Interior" (Chicago, IL: 2004).

55. Corless, "The Significance and Status of the Quincy Smelting Works, Franklin Township, Michigan (White Paper)," 7.

56. "Proposed Plan for Record of Decision Amendment for Operable Unit 3 at Torch Lake," ed. Region 5 U.S. EPA (Chicago: U.S. EPA, 2008), 4.

57. "Responsiveness Summary, Torch Lake Superfund Site, Operable Units I and II, Houghton County, Michigan," RS32.

58. Ibid., RS30-RS32.

59. Ibid., RS32-RS33.

60. Bode J. Morin, personal observation.

61. "Proposed Plan for Record of Decision Amendment for Operable Unit 3 at Torch Lake Superfund Site," 4.

62. "EPA Completes Construction at Torch Lake," 1 and 3. By 2004, the Lake Linden sands and Hubbell/Tamarack locations had been delisted from the NPL.

63. Bode J. Morin, personal involvement with the conference calls and follow up meetings.

64. "Letter from EPA Region 5 to Mr. Jim Corless, Superintendant, Keweenaw National Park," (2009).

65. Garrett Neese, "Smelter Group Seeks Money from Council," *Daily Mining Gazette*, September 25, 2008.

66. "Proposed Plan for Record of Decision Amendment for Operable Unit 3 at Torch Lake Superfund Site," 5.

67. "EPA Proposes Cleanup Plan for Quincy Smelter Erosion," 1.

68. Kurt Hauglie, "$1 Million Earmarked for Smelter," *Daily Mining Gazette*, June 27, 2009.

Conclusion

1. The most important copper mines operated in Vermont, Tennessee, Michigan, Colorado, Montana, Idaho, New Mexico, Utah, Arizona, Nevada, California, and Alaska.

2. "Copper Statistics and Information," USGS Minerals Information, accessed July 8, 2009, http://minerals.usgs.gov/minerals/pubs/commodity/copper/ and U. S. Geological Survey, "Copper Statistics," U.S. Geological Survey, accessed July 8, 2009, http://minerals.usgs.gov/ds/2005/140/copper.pdf. Despite average production over 2 million metric tons per year in the late 1990s, the average annual production of refined copper including scrap in the US, from 1908–2008, was 1.19 million metric tons. In 2008, Chile and Peru were the top two copper producers in the world.

Appendix

1. In reality, mine productivity and profits do not map to simple bell curves, but rather are significantly influenced by the fluctuating demand and price of the commodity, lessening ore quality, and evolving technology in complex and myriad ways, leading to many productivity peaks and valleys over the life of the mine. Mining companies, especially in the twentieth century, diversified production, opened new ore deposits, and vertically integrated to hedge against declining mine production in any one region.

2. "County Population Census Counts 1900–90." U.S. Census Bureau, accessed August 12, 2009,www.census.gov/population/www/censusdata/cencounts/index.html

3. This argument, however, is not intended to minimize the significant conflicts that erupted in mining districts between mine management and labor over issues of safety, wages, paternalism, and unionization as production increased and then decreased, or to suggest that political conflict did not disrupt the continuous development of certain regions, or that mining companies did not garner substantial and unlawful political control. It merely suggests that mining growth and expansion tended to be viewed broadly as providing positive economic returns. Although the identity of a region on a broad scale can be seen as positive with positive growth, often that positive identity is more clearly focused in later phases when the negative regional perceptions that accompany production declines, abandonment, and considerations of contamination cause reflection on earlier, better times.

4. Quoted in Chavez, "Does 'the company' owe the workers?" *Independent Record*, July 25, 1982, 8B.

Bibliography

Audio Visual Sources

Copper Basin Pictorial Calendar. Ducktown, TN: Copper Duck Press, 2008.

Mining the Berkeley Pit 1955–1982. (Interpretive Sign.) Butte, MT: Butte Chamber of Commerce, 2004 (est.).

Old Works Vision to Reality. Butte, MT: Maney Telefilms, ARCO Environmental Remediation, 1998 (est.). Videocassette (VHS), 13 minutes.

A Story of Rebirth. (Interpretive Sign.) Butte, MT: ARCO, 2004 (est.).

Books and Journal Articles

Agricola, Georgius. *De Re Metallica*. Translated by Herbert Hoover. New York: Dover Publications, 1950.

Aiken, Katherine. "Western Smelters and the Problem of Smelter Smoke." In *Northwest Lands, Northwest Peoples: Readings in Environmental History*, edited by Dale Goble and Paul W. Hirt. Seattle: University of Washington Press, 1999.

"Anaconda Copper Smelter May Switch to Coal." *Chemical Week* 123, (1980): 61.

Avrami, Erica, Randall Mason, and Marta de la Torre. *Values and Heritage Conservation*. Los Angeles, CA: The Getty Conservation Institute, 2000.

Bacha, Charles P., Joseph L. Schwalje, and Anthony J. Del Mastro. *Elements of Engineering Materials*. New York: Harper and Brothers, 1957.

Bakken, Gordon Morris. "Montana, Anaconda, and the Price of Pollution." *The Historian* 69, no. 1 (2007): 36–48.

Barclay, Robert Edward. *The Copper Basin, 1890 to 1963*. Knoxville, TN: R. E. Barclay, 1975.

——. *Ducktown Back in Raht's Time*. Chapel Hill: The University of North Carolina Press, 1946.

Barnhardt, Wilton. "The Death of Ducktown." *Discover* 8 (1987): 35–42.

Bass, George F. "The Cape Gelidonya Wreck: Preliminary Report." *American Journal of Archaeology* 65 (1961): 267–76.

Baum, Dan. "Butte, America-Poisoned, Ruined and Self-Cannibalized." *American Heritage Magazine* 48, no. 2 (1997): 57–67.

Baum, Dan, and Margaret L. Knox. "We Want People Who Have a Problem with Mine Wastes to Think of Butte." *Smithsonian* 23, no. 8 (1992): 46–55.

Bijker, Wiebe E., Thomas P. Hughes, and Trevor Pinch, eds. *The Social Construction of Technological Systems: New Directions in the Sociology and History of Technology*. Cambridge, MA: MIT Press, 1987.

Biringuccio, Vannoccio. *The Pirotechnia of Vannoccio Biringuccio*. Cambridge, MA: M.I.T. Press, 1966.

Biswas, A. K., and W. G. Davenport. *Extractive Metallurgy of Copper*. New York: Pergamon Press, 1980.

Brady, George S., and Henry R. Clauser. *Materials Handbook*. New York: McGraw-Hill, 1986.

Bray, John Leighton. *Non-Ferrous Production Metallurgy*. 2nd ed. New York: J. Wiley and Sons, 1953.

Bridges, E. M. *Healing the Scars: Derelict Land in Wales*. Swansea, Wales, 1988.

Burt, Roger. "The International Diffusion of Technology in the Early Modern Period: The Case of the British Non-Ferrous Mining Industry." *Economic History Review, New Series* 44, no. 2 (1991): 249–71.

——. "The Transformation of the Non-Ferrous Metals Industries in the Seventeenth Century and Eighteenth Centuries." *Economic History Review, New Series* 48, no. 1 (1995): 23–45.

Butts, Allison. *Copper: The Science and Technology of the Metal, Its Alloys, and Compounds*. New York: Hafner Publishing, 1970.

Campbell, Gary A. "A Retrospect of U.S. Oil Industry Takeovers of U.S. Copper Companies." *Materials and Society* 15, no. 2 (1991): 89–107.

Chaput, Donald. *The Cliff: America's First Great Copper Mine*. Kalamazoo, MI: Sequoia Press, 1971.

Childe, V. Gordon. "The Danube Thoroughfare and the Beginning of Civilization in Europe." *Antiquity* 1, no. 1 (1927): 79–91.

Christensen, Charles. "History of Converting." *The Mining World* 33 (1910): 1037–39.

Clay, Grady. "Copper Basin Cover-Up." *Landscape Architecture* 73, no. 4 (1983): 49–55, 94.

Cloverland Tourist Guide. Menomonee, MI: The Herald-Leader Company, 1926.

Coghlan, H. H. "Some Experiments on the Origin of Early Copper." *Man* 39 (July 1939): 106–108.

Conant, Henry D. "Copper Smelting in Michigan." *The School of Mines Quarterly* 42, no. 4 (1911): 285–97.

———. "Copper-Smelting Practice in Lake Superior Region." *Mining and Engineering World* 35 (1911): 294–296.

———. "The Historical Development of Smelting and Refining Native Copper." *Mining Congress Journal* 17 (1931): 531–32.

Cooper, James B. "Historical Sketch of Smelting and Refining Lake Copper." *Proceedings of the Lake Superior Mining Institute* 7 (1901): 44–50.

Copper: From Mine to Finished Product. New York: Anaconda Copper Mining Company, 1920.

Costagliola, Pilario, Marco Benvenuti, Laura Chiarantini, Sara Bianchi, Francesco Di Benedetto, Mario Paolieri, and Luca Rossato. "Impact of Ancient Metal Smelting on Arsenic Pollution in the Pecora River Valley, Southern Tuscany, Italy." *Applied Geochemistry* 23, no. 5 (2008): 1241–59.

Craddock, Paul T. "The Composition of Copper Alloys Used by the Greek, Etruscan, and Roman Civilizations. 2. The Archaic, Classical and Hellenistic Greeks." *Journal of Archaeological Science* 4 (1977): 103–33.

———. "The Composition of Copper Alloys Used by the Greek, Etruscan, and Roman Civilizations. 3. The Origins and Early Use of Brass." *Journal of Archaeological Science* 5 (1978): 1–16.

———. "The Composition of the Copper Alloys Used by the Greek, Etruscan, and Roman Civilizations. 1. The Greeks Before the Archaic Period." *Journal of Archaeological Science* 3 (1976): 93–113.

———. *Early Metal Mining and Production* Washington, DC: Smithsonian Institution Press, 1995.

Cravens, Curtis. *Copper on the Creek: Reclaiming an Industrial History*. New York: Place in History, 2000.

Cronon, William. *Changes in the Land: Indians, Colonists, and the Ecology of New England*. New York: Hill and Wang, 1983.

———. "Kennecott Journey: The Paths out of Town." In *Under an Open Sky: Rethinking America's Western Past*, 28–51. New York: W. W. Norton and Company, 1992.

———. *Nature's Metropolis: Chicago and the Great West*, edited by William Cronon. New York: W.W. Norton, 1991.

Cruikshank, Ken, and Nancy B. Bouchier. "Blighted Areas and Obnoxious Industries; Constructing Environmental Inequality on an Industrial Waterfront, Hamilton, Ontario, 1890–1960." *Environmental History* 9, no. 3 (2007): 30.

Culver, William W., and Cornel J. Reinhart. "Capitalist Dreams: Chile's Response to Nineteenth-Century World Copper Competition." *Comparative Studies in Society and History* 31, no. 4 (1989): 722–44.

Curtis, Kent. "Greening Anaconda." In *Beyond the Ruins: The Meanings of Deindustrialization*, edited by Jefferson Cowie and Joseph Heathcott, 91–111. Ithaca, NY: Cornell University Press, 2003.

Cushing, Frank Hamilton. "Primitive Copper Working: An Experimental Study." *American Anthropologist* 7, no. 1 (1894): 93–117.

Daniels, Karen. *Tennessee's Historic Copper Basin Area: An Overview*. Benton, TN: Polk County Publishing, 1992.

Darwin, Charles. *Journal of Researches into the Natural History and Geology of the Countries Visited During the Voyage Round the World of H.M.S. Beagle*. 11th ed. London: John Murray, Albemarle Street, 1913.

Davis, J. R. *ASM Specialty Handbook: Copper and Copper Alloys*. Materials Park, OH: ASM International, 2001.

Davis, Jr., R. A., A. T. Welty, J. Borrego, J. A. Morales, J. G. Pendon, and J. G. Ryan. "Rio Tinto Estuary (Spain): 5000 Years of Pollution." *Environmental Geology* 39, no. 10 (2000): 1107–1116.

Day, Joan, and R. F. Tylecote. *The Industrial Revolution in Metals*. London: The Institute of Metals, 1991.

Dayton, J. E. "The Problem of Tin in the Ancient World." *World Archaeology* 3, no. 1 (1971): 49–70.

"Detroit and Waterbury Copper Smelting Works." *The Mining Magazine* 1 (1853): 298.

Dobb, Edwin. "Pennies from Hell." *Harper's Magazine,* October 1996, 39–54.

Donald. Maxwell, Bruce. *Elizabethan Copper: The History of the Company of Mines Royal 1568–1605.* London: Pergamon Press, 1955.

Egleston, N. H. "Copper Refining in the United States." *Transactions of the American Institute of Mining Engineers* 9 (1881): 678–730.

Egleston, T. "The Point Shirley Works." *School of Mines Quarterly* 7 (1886): 360–84.

Ek, Anna S., Stefan Lofgren, Johan Bergholm, and Ulf Qvarfort. "Environmental Effects of One Thousand Years of Copper Production at Falun, Central Sweden." *Ambio* 30, no. 2 (2001): 96–103.

Emmons, David M. "The Price of 'Freedom': Montana in the Late and Post-Anaconda Era." *Montana: The Magazine of Western History* 44, no. 4 (1994): 66–73.

Fairlie, Andrew Miller. *Sulfuric Acid Manufacture.* New York: Reinhold Pub. Corp., 1936.

Farmer, Silas. *The History of Detroit and Michigan; or, the Metropolis Illustrated; a Full Record of Territorial Days in Michigan, and the Annals of Wayne County.* 2nd ed. Detroit, MI: S. Farmer & Co., 1889.

Finn, Janet L. *Tracing the Veins: Of Copper, Culture, and Community from Butte to Chuquicamata.* Berkeley: University of California Press, 1998.

Francaviglia, Richard V. *Hard Places: Reading the Landscape of America's Historic Mining Districts.* The American Land and Life Series. Iowa City: University of Iowa Press, 1991.

Gabb, Gerald. *Lower Swansea Valley Factsheet 7: The Rise and Fall of the Copper Industry.* Swansea: Swansea Museum's Service, 1990 (est.).

Gates, William Bryan. *Michigan Copper and Boston Dollars: An Economic History of the Michigan Copper Mining Industry.* Cambridge, MA: Harvard University Press, 1951.

Gillispie, Charles C., ed. *Diderot Pictorial Encyclopedia of Trades and Industry: Manufacturing and the Technical Arts in Plates, Selected from L'encyclopédie; Ou, Dictionnaire Raisonné Des Sciences, Des Arts Et Des Métiers, of Denis Diderot.* New York: Dover Publications, 1959.

Gladstone, J. H. "On the Transition from the Use of Copper to That of Bronze." *The Journal of the Anthropological Institute of Great Britain and Ireland* 26 (1897): 309–320.

Goin, Peter, and Elizabeth Raymond. "Living in Anthracite: Mining Landscape and Sense of Place in Wyoming Valley, Pennsylvania." *The Public Historian* 23, no. 2 (2001): 29–45.

Gordon, Robert B., and Patrick M. Malone. *The Texture of Industry: An Archaeological View of the Industrialization of North America.* New York: Oxford University Press, 1994.

Gorman, Hugh. "Manufacturing Brownfields: The Case of Neville Island, Pennsylvania." *Technology and Culture* 38, no. 3 (1997): 36.

Gowland, William. "Copper and Its Alloys in Early Times." *Journal of the Institute of Metals* 7 (1912): 23–49.

———. "The Metals in Antiquity." *Journal of the Royal Anthropological Institute of Great Britain and Ireland* 42 (1912): 235–287.

———. "Presidential Address: Copper and Its Alloys in Prehistoric Times." *The Journal of the Anthropological Institute of Great Britain and Ireland* 36 (Jan –June 1906): 11–38.

Grattan, J. P., D. D. Gilbertson, and C. O. Hunt. "The Local and Global Dimensions of Metalliferous Pollution Derived from a Reconstruction of an Eight Thousand Year Record of Copper Smelting and Mining at a Desert-Mountain Frontier in Southern Jordan." *Journal of Archaeological Science* 34, no. 1 (2007): 83–110.

Grattan, J. P., G. K. Gillmore, D. D. Gilbertson, F. B. Pyatt, C. O. Hunt, S. J. McLaren, P. S. Phillips, and A. Denman. "Radon and 'King Solomon's Miners': Faynan Orefield, Jordanian Desert." *Science of The Total Environment* 319, no. 1–3 (2004): 99–113.

Greenberg, Peter. *Don't Go There!: The Travel Detective's Essential Guide to the Must-Miss Places of the World.* Emmaus, PA: Rodale, 2009.

Greenwood, William Henry. *A Manual of Metallurgy.* New York: G. P. Putnam's Sons, 1875.

Griffiths, Ralph A. *The City of Swansea: Challenges and Change.* Phoenix Mill, UK: Alan Stutton, 1990.

Hall, B. Clarence, and C. T. Wood. *The South.* New York: Scribner, 1995.

Hammersley, G. "The Effect of Technical Change in the British Copper Industry between the Sixteenth and Eighteenth Centuries." *Journal of European Economic History* 20 (1991): 155–73.

———. "Technique or Economy; the Rise and Decline of the Early English Copper Industry." *Business History* 15, no. 1 (1973): 1–31.

Hammett, Dashiell. *Red Harvest*, New York: Vintage Books, 1972.

Hardesty, Donald. "Issues in Preserving Toxic Wastes as Heritage Sites." *The Public Historian* 23, no. 2 (2001): 19–28.

Harris, J. R. "Copper and Shipping in the Eighteenth Century." *Economic History Review* 19, no. 3 (1966): 550–68.

Hedgcock, George Grant. "Injury by Smelter Smoke in Southwestern Tennessee." *Journal of the Washington Academy of Sciences* 4, no. 4 (1914): 70–71.

Herbert, Eugenia W. *Red Gold of Africa: Copper in Pre-colonial History and Culture*. Madison: University of Wisconsin, 1984.

Hofman, H. O. *Metallurgy of Copper*. New York: McGraw-Hill, 1924.

Hong, S., J. P. Candelone, C. C. Patterson, and C. F. Boutron. "History of Ancient Copper Smelting Pollution During Roman and Medieval Times Recorded in Greenland Ice." *Science* 272, no. 5259 (1996): 246–49.

Hong, S., J. P. Candelone, M. Soutif, and C. F. Boutron. "A Reconstruction of Changes in Copper Production and Copper Emissions to the Atmosphere During the Past 7000 Years." *The Science of the Total Environment* 188, no. 2-3 (1996): 183–95.

Hong, Sungmin, Jean-Pierre Candelone, Clair C. Patterson, and Claude F. Boutron. "Greenland Ice Evidence of Hemispheric Lead Pollution Two Millennia Ago by Greek and Roman Civilizations." *Science* 265, no. 5180 (1994): 1841–43.

Howard, Peter. *Heritage: Management, Interpretation, Identity*. New York: Continuum, 2003.

Hughes, Stephen. *Copperopolis: Landscapes of the Early Industrial Period in Swansea*. Aberystwyth, Ceredigion: Royal Commission on the Ancient and Historical Monuments of Wales, 2000.

Hunt, Chris O., David D. Gilbertson, and Hwedi A. El-Rishi. "An 8000-Year History of Landscape, Climate, and Copper Exploitation in the Middle East: The Wadi Faynan and the Wadi Dana National Reserve in Southern Jordan." *Journal of Archaeological Science* 34, no. 8 (2007): 1306–38.

Hyde, Charles K. *Copper for America: The United States Copper Industry from Colonial Times to the 1990s*. Tucson: University of Arizona Press, 1998.

Jenkins, R. "Copper Smelting in England: Revival at the End of the Seventeenth Century." *Transactions of the Newcomen Society* 24 (1943): 73–80.

———. "The Reverberatory Furnace with Coal Fuel." *Transactions of the Newcomen Society* 14 (1934): 67–81.

Johnson, Edward A. "Industrial Destruction Reversed at Copper Basin." *National Woodlands* 25, no. 2 (2002): 10–13, 29.

King, Thomas. *Cultural Resource Laws and Practice.* Walnut Creek, CA: Altamira Press, 2008.

Kuzell, Charles. "The Development of Modern Copper Smelting." *Transactions of the Metallurgical Society of AIME* 218 (August 1960): 578–84.

Laist, Frederick. "Seventy-Five Years of Progress in Smelting and Leaching of Ores." In *Seventy-Five Years of Progress in the Mineral Industry, 1871–1946,* 126–61. New York: The American Institute of Mining and Metallurgical Engineers, 1948.

Lankton, Larry D. *Beyond the Boundaries: Life and Landscape at the Lake Superior Copper Mines, 1840–1875.* New York: Oxford University Press, 1997.

———. *Cradle to Grave: Life, Work, and Death at the Lake Superior Copper Mines.* New York: Oxford University Press, 1991.

———. *Hallowed Ground: Copper Mining and Community Building on Lake Superior, 1840s–1990s (Draft Version).* Detroit, MI: Wayne State University Press, 2010.

Lankton, Larry D., and Charles K. Hyde. *Old Reliable: An Illustrated History of the Quincy Mining Company.* Hancock, MI: Quincy Mine Hoist Association, 1982.

Leitner, Jonathan. "The 1869 Copper Tariff: The Politics and Geography of U.S. Development in World-Systems Perspective." *Sociological Perspectives* 43, no. 3 (2000): 437–97.

———. "Red Metal in the Age of Capital: The Political Ecology of Copper in the Nineteenth-Century World-Economy." *Review-Fernand Braudel Center for the Study of Economies, Historical Systems, and Civilizations* 24, no. 3 (2001): 373–439.

Lock, Alfred G., and Charles G. Lock. *A Practical Treatise on the Manufacture of Sulphuric Acid.* London: Sampson, Low, Marston, Searle, and Rivington, 1879.

Lovell, Endicott R., and Herman C. Kenny. "Present Smelting Practice." *Mining Congress Journal* 17 (1931): 533–38.

Lowenthal, David. *The Heritage Crusade and the Spoils of History.* New York: Cambridge University Press, 1998.

———. *The Past Is a Foreign Country.* New York: Cambridge University Press, 1985.

———. "Stewarding the Past in a Perplexing Present." In Avrami et al. *Values and Heritage Conservation.* Los Angeles, CA: The Getty Conservation Institute, 2000.

Lyman, Taylor. *Metals Handbook.* Novelty, OH: American Society for Metals, 1961.

Lynch, Martin. *Mining in World History.* London: Reaktion, 2002.

Makra, L., and P. Brimblecombe. "Selections from the History of Environmental Pollution, with Special Attention to Air Pollution. Part 1." *International Journal of Environment and Pollution* 22, no. 6 (2004): 641–656.

Malone, Michael P. *The Battle for Butte: Mining and Politics on the Northern Frontier, 1864–1906.* Seattle: University of Washington Press, 1981.

———. "The Close of the Copper Century." *Montana: The Magazine of Western History* 35, no. 2 (1985): 69–72.

Malone, Michael P., Richard B. Roeder, and William L. Lang. *Montana: A History of Two Centuries.* Rev. ed. Seattle: University of Washington Press, 1991.

Manasse, Andreal, and Marcello Mellini. "Chemical and Textural Characterisation of Medieval Slags from the Massa Marittima Smelting Sites (Tuscany, Italy)." *Journal of Cultural Heritage* 3 (2002): 187–206.

Marcosson, Isaac Frederick. *Anaconda.* New York: Dodd, Mead & Company, 1957.

———. *Copper Heritage; The Story of Revere Copper and Brass Incorporated.* New York: Dodd, Mead, 1955.

Marcus, Jerrold J. "Butte: Richest Hill on Earth and Costliest Mine Superfund Site." *Engineering and Mining Journal* 201, no. 2 (2000): 31–44.

Martin, Susan R. *Wonderful Power: The Story of Ancient Copper Working in the Lake Superior Basin.* Detroit, MI: Wayne State University Press, 1999.

Mattson, Margaret E., and Tee L. Guidotti. "Health Risks Associated with Residence near a Primary Copper Smelter: A Preliminary Report." *American Journal of Industrial Medicine* 1 (1980): 365–74.

Maysilles, Duncan. *Ducktown Smoke: The Fight over One of the South's Greatest Environmental Disasters.* Chapel Hill: University of North Carolina Press, 2011.

McMillan, Donald. *Smoke Wars Anaconda Copper, Montana Air Pollution, and the Courts, 1890–1924.* Helena: Montana Historical Society, 2000.

Melosi, Martin V., and Philip Scarpino, eds. *Public History and the Environment.* Malabar, FL: Krieger Publishing Company, 2004.

Mercier, Laurie. *Anaconda: Labor, Community, and Culture in Montana's Smelter City.* Urbana: University of Illinois Press, 2001.

Miskimin, Harry A. *The Economy of Later Renaissance Europe.* Cambridge, U.K.: Cambridge University Press, 1977.

Morag-Levine, Noga. *Chasing the Wind: Regulating Air Pollution in the Common Law State.* Princeton, NJ: Princeton University Press, 2003.

Morral, F. R. "A Mini-History of the Rio Tinto (Spain) Region." *CIM Bulletin* 83, no. 935 (1990): 5.

Napier, James. "On Copper Smelting." *The Philosophical Magazine and Journal of Science* 4 (1852): Non-consecutive pages, 45–59, 262–71, 345–55, 453–65.

———. "On Copper Smelting." *The Philosophical Magazine and Journal of Science* 5 (1853): Non-consecutive pages, 30–39, 175–84, 345–54, 486–93.

Newell, Edmund. "Atmospheric Pollution and the British Copper Industry, 1690–1920." *Technology and Culture* 38, no. 3 (1997): 655–89.

———. "Copperopolis: The Rise and Fall of the Copper Industry in Swansea District." *Business History* 32, no. 3 (1990): 75–97.

Newell, Edmund, and Simon Watts. "The Environmental Impact of Industrialization in South Wales in the Nineteenth Century: 'Copper Smoke' and the Llanell." *Environment and History* 2, no. 3 (1996): 309–36.

Nriagu, Jerome O. *Copper in the Environment.* New York: Wiley, 1979.

———. "Global Metal Pollution: Poisoning the Biosphere." *Environment* 32, no. 7 (1990): 7–33.

———. *Lead and Lead Poisoning in Antiquity.* New York: Wiley, 1983.

Olson, James M. *Michigan Environmental Law: A Guide to Environmental Law.* Traverse City, MI: Neahtawanta Press, 1981.

O'Neill, Hugh. "The Inventions of Bessemer in Relation to Non-Ferrous Metals." *Metallurgia* 54, no. 326 (1956): 269–73.

Ortega, Luis. "The First Four Decades of the Chilean Coal Mining Industry, 1840–1879." *Journal of Latin American Studies* 14, no. 1 (1982): 1–32.

Patterson, Clair C. "Native Copper, Silver, and Gold Accessible to Early Metallurgists." *American Antiquity* 36, no. 3 (1971): 286–321.

Percy, John. *Metallurgy. The Art of Extracting Metals from Their Ores.* London: J. Murray, 1861.

Peters, Edward Dyer. *Modern Copper Smelting.* New York: The Scientific Publishing Co., 1898.

Pettengill, Robert B. "The United States Foreign Trade in Copper." *American Economic Review* 25, no. 3 (1935): 426–41.

Piggot, A. Snowden. *The Chemistry and Metallurgy of Copper, Including a Description of the Principal Copper Mines of the United States and Other Countries, the Art of Mining and Preparing Ores for Market, and the Various Processes of Copper Smelting.* Philadelphia, PA: Lindsay & Blakiston, 1858.

Prain, Ronald. *Copper: The Anatomy of an Industry.* London: Mining Journal Books, 1975.

Price, Jon. "Interpreting Industrial Heritage." In *Heritage Interpretation*, edited by Allison Helms and Marion Blockley, 111–122. New York: Routledge, 2006.

Przeworski, Joanne Fox. *The Decline of the Copper Industry in Chile and the Entrance of North American Capital: 1870–1916.* New York: Arno Press, 1980.

Pursell, Carrol. "The Rise and Fall of the Appropriate Technology Movement in the United States, 1965–1985." *Technology and Culture* 34, no. 3 (1993): 629–37.

Pyatt, F. B. "Copper and Lead Bioaccumulation by Acacia Retinoides and Eucalyptus Torquata in Sites Contaminated as a Consequence of Extensive Ancient Mining Activities in Cyprus." *Ecotoxicology and Environmental Safety* 50, no. 1 (2001): 60–64.

Pyatt, F. B., G. W. Barker, P. Birch, D. D. Gilbertson, J. P. Grattan, and D. J. Mattingly. "King Solomon's Miners—Starvation and Bioaccumulation? An Environmental Archaeological Investigation in Southern Jordan." *Ecotoxicology and Environmental Safety* 43, no. 3 (1999): 305–08.

Pyatt, F. B., G. Gilmore, J. P. Grattan, C. O. Hunt, and S. McLaren. "An Imperial Legacy? An Exploration of the Environmental Impact of Ancient Metal Mining and Smelting in Southern Jordan." *Journal of Archaeological Science* 27, no. 9 (2000): 771–78.

Pyatt, F. B., and J. P. Grattan. "Some Consequences of Ancient Mining Activities on the Health of Ancient and Modern Human Populations." *Journal of Public Health* 23, no. 3 (2001): 235–236.

Pyatt, F. B., A. J. Pyatt, C. Walker, T. Sheen, and J. P. Grattan. "The Heavy Metal Content of Skeletons from an Ancient Metalliferous Polluted Area in Southern Jordan with Particular Reference to Bioaccumulation and Human Health." *Ecotoxicology and Environmental Safety* 60, no. 3 (2005): 295–300.

Quinn, M. -L. "The Appalachian Mountains' Copper Basin and the Concept of Environmental Susceptibility." *Environmental Management* 15, no. 2 (1991): 179–94.

———. "Early Smelter Sites: A Neglected Chapter in the History of and Geography of Acid Rain in the United States." *Atmospheric Environment* 23, no. 6 (1989): 1281–92.

———. "Industry and Environment in the Appalachian Copper Basin, 1890–1930." *Technology and Culture* 34, no. 3 (1993): 575–612.

———. "Should All Degraded Landscapes Be Restored? A Look at the Appalachian Copper Basin." *Land Degradation & Rehabilitation* 3 (1992): 115–134.

Quivik, Fredric L. "The Anaconda Company Smelters: Great Falls and Anaconda." *The Speculator* 1, no. 2 (1984): 38–45.

———. "Integrating the Preservation of Cultural Resources with Remediation of Hazardous Materials: An Assessment of Superfund's Record." *The Public Historian* 23, no. 2 (2001): 47–61.

———. "Landscapes as Industrial Artifacts: Lessons from Environmental History." *IA: The Journal of the Society for Industrial Archeology* 26, no. 2 (2000): 55–64.

———. "Of Tailings, Superfund Litigation, and Historians as Experts: U.S. V. ARCO, Et Al. (the Bunker Hill Case in Idaho)." *The Public Historian* 26, no. 1 (2004): 81–104.

Raymond, Robert. *Out of the Fiery Furnace: The Impact of Metals on the History of Mankind.* University Park: Pennsylvania State University Press, 1986.

Rehren, Thilo. "Small Size, Large Scale Roman Brass Production in Germania Inferior." *Journal of Archaeological Science* 26, no. 8 (1999): 1083–87.

Richter, F. E. "The Copper-Mining Industry in the United States, 1845–1925." *Quarterly Journal of Economics* 41, no. 2 (1927): 236–91.

Robertson, David. "Heaps of History: Toluca and the Historic Longwall Mining District." *Journal of Illinois History* 3 (Autumn 2000): 162–84.

Rothenberg, Benno, H. G. Bachmann, and Studies Institute for Archaeo-Metallurgical. *The Ancient Metallurgy of Copper: Archaeology-Experiment-Theory*, Metal in History 3. London: Institute for Archaeo-Metallurgical Studies, 1990.

Scheinberg, I. Herbert. "Human Health Effects of Copper." In *Copper in the Environment*, edited by Jerome O. Nriagu, 17–33. New York: John Wiley and Sons, 1979.

Schmitz, Christopher J. *World Non-Ferrous Metal Production and Prices 1700–1976*. London: Frank Cass, 1979.

Schnabel, Carl. *Handbook of Metallurgy*. New York: Macmillan, 1898.

Seigworth, Kenneth J. "Ducktown-a Postwar Challenge: Can Man Correct a Century of Land Abuse in Tennessee's Copper Basin? A Start Has Been Made." *American Forests* 49 (November 1943): 521–23, 58.

Shaler, Nathaniel Southgate. *Man and the Earth.* New York: Fox, Duffield & Company, 1905.

Shapiro, Aaron. "Air Conditioned by the Cool Breezes of Lake Superior: Vacationing in Michigan's Copper Country." In *New Perspectives on Michigan's Copper Country*, edited by Alison K. Hoagland, Erik C. Nordberg, and Terry S. Reynolds, 135–153. Hancock, MI: Quincy Mine Hoist Association, 2007.

Shovers, Brian. "Remaking the Wide-Open Town: Butte at the End of the Twentieth Century." *Montana: The Magazine of Western History* 48, no. 3 (1998): 40–53.

Shovers, Brian, Mark Fiege, Dale Martin, and Fred Quivik. *Butte & Anaconda Revisited: An Overview of Early-Day Mining and Smelting in Montana.* Special Publication 99. Butte, MT: Montana Bureau of Mines and Geology; Montana College of Mineral Science and Technology, 1991.

Slack, Jennifer Daryl. "The Politics of the Pristine." *Topia* 2 (1998): 24.

Smith, Duane A. *Mining America: The Industry and the Environment. 1800–1980.* Lawrence: University of Kansas Press, 1987.

Smith, Pamela H., and Paula Findlen. *Merchants & Marvels: Commerce, Science, and Art in Early Modern Europe.* New York: Routledge, 2002.

St. Clair, Jeffery. "Something About Butte." In *Red State Rebels: Tales of Grassroots Resistance in the Heartland* edited by Joshua Frank and Jeffery St. Clair, 90–102. Oakland, CA: AK Press, 2008.

Stevens, Horace J. *The Copper Handbook*, vol. 10. Houghton, MI: Horace J. Stevens, 1911.

Sticht, Robert. "Pyritic Smelting-Its History, Principals, Scope, Apparatus, and Practical Results." In *Modern Copper Smelting*, edited by Edward Dyer Peters, 396–441. New York: The Scientific Publishing, Co., 1898.

Stine, Jeffery K., and A. Tarr Joel. "At the Intersection of Histories: Technology and the Environment." *Technology and Culture* 39, no. 4 (1998): 601–640.

Strom, H. C. "Parallel between the British and the Continental Methods of Copper Smelting." *The Mining Review* 7 (July 1835): 66–13.

Swansea, City and County of. *A Short History of the Hafod Works*: Swansea, Wales: Welsh Assembly Government, 2007.

Theophilus. *On Divers Arts; the Treatise of Theophilus.* Translated by John G. Hawthorne and Cyril Stanley Smith. Dover Publications, Inc. (1979 reprint ed.) Chicago: University of Chicago Press, 1963.

Tidball, Eugene C. "What Ever Happened to the Anaconda Company?" *Montana: The Magazine of Western History* 47, no. 2 (1997): 60–68.

"Traveling Notes in South Wales." *The Mirror of Literature, Amusement, and Instruction* 17, no. 469 (1831): 1.

Toomey, Robert R. *Vivian and Sons, 1809–1924: A Study of the Firm in the Copper and Related Industries*, British Economic History. New York: Garland, 1985.

Tylecote, R. F. *A History of Metallurgy.* London: Institute of Metals, 1976.

———. *A History of Metallurgy.* 2nd ed. The Institute of Materials, 1992.

Ulke, Titus. "Notes on the History of Electrolytic Copper Refining in America." *Electrochemical Industry* 1, no. 7 (1902–3): 240.

Underwood, E. Ashworth. "The History of Cholera in Great Britain." Proceedings of the Royal Society of Medicine, XLI, 1947, 165.

Valenzuela, Luis "The Chilean Copper Smelting Industry in the Mid-Nineteenth Century: Phases of Expansion and Stagnation, 1834–58." *Journal of Latin American Studies* 24, no. 2 (1992): 507–50.

Vivian, John Henry. "An Account of the Process of Smelting Copper as Conducted at the Hafod Works, near Swansea." *The Annals of Philosophy* 5 (January –June 1823): 113–24.

Wallace, Mike. *Mickey Mouse History and Other Essays on American Memory.* Philadelphia, PA: Temple University Press, 1996.

Walters, Henry Beauchamp, and Samuel Birch. *History of Ancient Pottery, Greek, Etruscan, and Roman.* 2 vols. London: J. Murray, 1905.

Warren, Kenneth. *Wealth, Waste, and Alienation: Growth and Decline in the Connellsville Coke Industry.* Pittsburgh, PA: University of Pittsburgh Press, 2001.

White, Richard. *The Organic Machine.* New York: Hill and Wang, 1995.

Wilson, Andrew. "Machines, Power, and the Ancient Economy." *The Journal of Roman Studies* 92 (2002): 1–31.

Wilson, Bob, Brennan Lang, and F. Brian Pyatt. "The Dispersion of Heavy Metals in the Vicinity of Britannia Mine, British Columbia, Canada." *Ecotoxicology and Environmental Safety* 60, no. 3 (2005): 269–76.

Wyckoff, William. "Postindustrial Butte." *Geographical Review* 85, no. 4 (1996): 18.

Young, Gordon, and Emory Kristoff. "The Tennessee Valley Today." *National Geographic* 143, no. 6 (1973): 830–863.

Young, Otis E. "The American Copper Frontier, 1640–1893." *The Speculator* 1, no. 2 (1984): 4–15.

———. "Origins of the American Copper Industry." *Journal of the Early Republic* 3, no. 2 (1983): 117–37.

Legal Rulings, Cases, and Consent Decrees

City and County of Butte-Silver Bow. (1985) "Council Bill No. 238, Ordinance 238 (Preservation Ordinance)," Butte, MT: Butte-Silver Bow.

Fields, T. (2001) "Memorandum of Understanding between Tennessee Department of Environment and Conservation, U.S. EPA and OXY, Inc.," *CECRLA Docket No. 01-10-C,* Atlanta, GA: U.S. EPA Region 4.

Hamilton, M. H. (2001) "Commissioner's Order," *Site ID 70-510, Case No. 99-0480,* Superfund. Nashville, TN: State of Tennessee, Department of Environment and Conservation.

———. (2001) "North Potato Creek VOAP Order," *Site ID 70-508, Case No. 99-0480, S*uperfund. Nashville: State of Tennessee, Department of Environment and Conservation.

Hankinson, J. H. (2001) "Administrative Order on Consent for Partial Payment," *CERCLA Docket No. 01-13-C,* edited by U.S. EPA. Atlanta, GA: U.S. EPA

———. (2001) "Administrative Order on Consent for Removal Action at the North Potato Creek Watershed," *CERCLA Docket No. 01-11-C,* edited by U.S. EPA. Atlanta, GA: U.S. EPA.

———. (2001) "Administrative Order on Consent for Removal Action at Davis Mill Creek Watershed," *CERCLA Docket No. 01-12-C,* edited by R. U.S. EPA. Atlanta, GA: U.S. EPA.

Supreme Court of Tennessee. (1903) "A.W. Swain v. Tennessee Copper Co. et al. and Thomas L. Cole v. Ducktown Sulphur, Copper & Iron Co. Limited, et al.," Knoxville, TN: Supreme Court of Tennessee.

United States Congress. (1966) "National Historic Preservation Act of 1966," *Public Law 102-575.* Washington, DC: U.S. Congress.

United States Supreme Court. (1907) "State of Georgia v. Tennessee Copper Company and Ducktown Sulphur, Copper, & Iron Company," Washington, DC: U.S. Supreme Court.

United States District Court for the District of Montana. (1998) "State of Montana v. Atlantic Richfield Company, Consent Decree," Helena, MT: United States District Court.

———. (1998) "United States v. Atlantic Richfield Company, State of Montana v. Atlantic Richfield Company, Streamside Tailings Operable Unit and Federal and Tribal Natural Resource Damages Consent Decree," Helena, MT: United States District Court.

United States Environmental Protection Agency. (1990) "Agreement and Covenant Not to Sue (Boliden Intertrade)," Washington, DC: U.S. EPA.

Newspaper Articles

Chattanooga Sunday Times (Chattanooga, TN)

Chattanooga Times Free Press (Chattanooga, TN)

Cherokee Scout (Cherokee, NC)

Copper City Advance (Copperhill, TN)

Daily Mining Gazette (Houghton, MI)

Detroit News (Detroit, MI)

Detroit Post and Tribune (Detroit, MI)

Great Falls Tribune (Great Falls, MT)

High Country News (Denver, CO)

Independent Record (Helena, MT)

Missoulian (Missoula, MT)

Montana Eagle (Billings, MT)

Montana Standard (Butte, MT)

New York Times (New York, NY)

Spokesman Review (Spokane, WA)

Adams, Duncan. "Did Toxic Stew Cook the Goose?" *High Country News*, December 11, 1995.

"Arcadian Mining Property Once the Rival of C&H." *Daily Mining Gazette*, June 24, 1978.

"Browning Is Speaker Saturday at Conservation League Meet." *Copper City Advance*, May 25, 1950.

Chavez, Lynda. "When Arco Left Town." *New York Times*, July 25, 1982.

———. "Does 'the company' owe the workers?" *Independent Record*, July 25, 1882.

Clayton, Ron. "Ducktown Basin Museum Hits 30 with Plans to Keep Growing." *Chattanooga Times Free Press*, October 13, 2008.

"The Detroit and Lake Superior Copper Co.'s Smelting Works." *Detroit Post and Tribune Newspaper*, 1879. In *Michigan Pioneer and Historical Collection* 28 (1898): 647–652.

Devlin, Sherry "Arco Recommends Leaving Milltown Sediment." *Missoulian*, November 6, 2001.

Dixon, Chris. "Up from the Mines in Tennessee." *New York Times*, December 14, 2007.

"Editor of 1872 Writes About Ducktown Mines." *Cherokee Scout*, September 5, 1935.

Egan, Timothy. "In Its Own Decay, Butte Sees a National Treasure." *New York Times*, August 30, 1997.

"EPA Chief Says Closure a Surprise." *The Montana Standard*, October 1, 1980.

Haffey, Vera. "Anacondan's Remember Day the Smelter Closed." *Montana Standard*, September 29, 2005.

Haugland, Keith. "Stubborn Stack Dies Slowly." *Great Falls Tribune*, September 19, 1982.

Hauglie, Kurt. "$1 Million Earmarked for Smelter." *Daily Mining Gazette*, June 27, 2009.

———. "EPA to Host Smelter Hearing." *Daily Mining Gazette*, January 12, 2009.

Lurgio, Jeremy. "Montanans Weigh Options on a Toxic Legacy." *New York Times*, May 7, 2001.

McCarthy, Bob. "Arco's Lethal Legacy." *Montana Eagle*, 1981.

———. "Lethal Legacy, Part II." *Montana Eagle*, December 9, 1981.

Meets, Shari. "Anaconda Co. Closes Smelter." *Montana Standard*, September 30, 1980.

"'Must Conserve Resources If We Survive,' - Browning." *Copper City Advance*, June 1, 1950.

Neese, Garrett. "Smelter Group Seeks Money from Council." *Daily Mining Gazette*, September 25, 2008.

Peters, Mouzon. "Healing the Scars in the Copper Basin." *Chattanooga Sunday Times*, September 2, 1956.

Robbins, Jim. "Butte Breaks Ground to Mop up a World-Class Mess." *New York Times*, July 21, 1998.

———. "In a Town Called Opportunity, Distress over a Dump." *New York Times*, August 24, 2005.

———. "Town Pins Hopes on Superfund Site." *New York Times*, April 6, 1997.

Steele, Karen Dorn. "Superfund Revived Butte." *Spokesman Review*, July 28, 2002.

van Sweringen, Hugh. "Shutdown Price Tag Will Be $40 Million Annually." *The Montana Standard*, September 30, 1980.

Williams, Florence. "Butte, Montana, Seeks a New Life." *High Country News*, November 29, 1993.

"Wrought in Fire: Story of the Old Waterbury Copper Mill." *Detroit News*, March 29, 1896.

Personal Communications

Bode J. Morin. Personal communication with the Ducktown Mayor. Ducktown, TN. October 22, 2008.

Bode J. Morin. Personal communication with federal official close to the cleanup. Ducktown, TN. October 23, 2008.

Bode J. Morin. Personal communication with Ducktown Basin Museum director. Ducktown, TN. October 24, 2008.

Bode J. Morin. Personal communication with a person close to the clean up who wished for anonymity. Ducktown, TN. October 22, 2008.

Bode J. Morin. Personal communication with the Butte-Silver Bow archivist. Butte, MT November 11, 2008.

Bode J. Morin. Personal communication with the state library manager. Helena, MT. November 17, 2008.

Bode J. Morin. Personal communication with the Butte-Silver Bow Planner. Butte, MT. November 20, 2008.

Bode J. Morin. Personal communication with Butte-Silver Bow archive staff. Butte, MT November 11, 2008.

Bode J. Morin. Personal communication with EPA official. Butte, MT. November 11, 2008.

Bode J. Morin. Personal communication with Anaconda-Deer Lodge Planner. Anaconda, MT. November 17, 2008.

Bode J. Morin. Personal communication with members of the Preserve the Stack Committee. Anaconda, MT. November 20, 2008.

Reports and Government Documents

Avrami, Erica, Randall Mason, and Marta de la Torre, eds. "Values and Heritage Conservation." Los Angeles, CA: Getty Conservation Institute, 2000.

Australia State Party. "The Burra Charter." International Charter for the Conservation and Restoration of Monuments and Sites (ICOMOS), 1999.

Butte-Silver Bow Local Government Study Commission, ed. "Minutes of the Meeting, May 26, 2005." Butte, MT: Butte-Silver Bow Local Government, 2005.

Camp Dresser and McKee, ed. "Second Five-Year Review Report for Silver Bow Creek/Butte Area Superfund Site." Helena, MT: U.S. EPA Region 8, 2005.

CH2M Hill, ed. "Final Work Plan for Anaconda Smelter RI/FS." Anaconda, MT: U.S. EPA, 1988.

"Cities of Houghton and Hancock City Councils Joint Resolution on Restoration of the Quincy Smelter (May 30, 2007)." Houghton and Hancock, MI: Houghton and Hancock City Councils, 2007.

Cities Service Company, ed. "Engineering Report on Land Reclamation: Copper Basin - Polk County, Tennessee." Copperhill, TN: Cities Service Company, 1973.

———. "Engineering Report: Proposed Treatment Facilities, Copperhill Plants." Copperhill, TN: Cities Service Company, 1973.

Community Involvement Partnership, ed. "Community Involvement Plan for the Copper Basin Mining District." Polk County, TN: U.S. EPA, Tennessee Department for Environment and Conservation, and Glenn Springs Holdings, Inc., 2001.

Daniels, Karen. "National Register of Historic Places Registration Form: Central Headframe." Edited by National Park Service. Washington, DC: Department of the Interior, 1992.

———. "National Register of Historic Places Registration Form: Copperhill Historic District." Edited by National Park Service. Washington, DC: Department of the Interior, 1992.

———. "National Register of Historic Places Registration Form: Ducktown Historic District." Edited by National Park Service. Washington, DC: Department of the Interior, 1992.

———. "National Register of Historic Places, Multiple Property Documentation Form: Historic Resources of the Tennessee Copper Basin," edited by National Park Service. Washington, DC: Department of the Interior, 1992.

Davis, Christine E. "Archaeological Survey and Land-Use History of The Pittsburgh Technological Center Site, Pittsburgh, Pennsylvania." Pittsburgh, PA: Urban Redevelopment Authority of Pittsburgh, 1989.

De la Torre, Marta, and Randall Mason, eds. "Assessing the Values of Cultural Heritage." Los Angeles: Getty Conservation Institute, 2002.

Dewey, Frederic P. "Bulletin of the United States National Museum, No. 42: A Preliminary Descriptive Catalogue of the Systematic Collections in Economic Geology and Metallurgy in the U. S. National Museum." Washington, DC: Smithsonian Institution, 1891.

Donohue and Associates, eds. "Final Remedial Investigation Report, Operable Unit III." Chicago, IL: U.S. EPA Region 5, 1992.

Dosch, Donald F. "National Register of Historic Places Inventory - Nomination Form for Butte Historic District," edited by National Park Service. Washington, DC: U.S. Department of the Interior, 1972.

Emmons, David M. "United States District Court, District of Montana, Butte Division; United States of America v. Atlantic Richfield Company, et al; Expert Report of David Emmons, Ph.D." Butte, MT: U.S. District Court, 1997.

Fiege, Mark, Fredric L. Quivik, and Brian Shovers. "Industrial Heritage of Butte and Anaconda: An Analysis of the Historical Significance of the Surviving Physical Features of the Anaconda Copper Mining Company." Butte, MT: Butte Historical Society, 1985.

"General Work Plan, Anaconda Smelter NPL Site, Attachment I." Anaconda, MT: U.S. EPA Region 8, 1988.

Heritage Conservation and Recreation Service, ed. "Butte, Montana: A Project Report." Butte, MT: National Park Service, 1980.

Hyde, Charles K. "Historic Report on the Quincy Mining Company," edited by National Park Service Washington, DC: Department of Interior, 1978.

ICOMOS. "Evaluations of Cultural Properties," edited by Bureau of Cultural Properties World Heritage Convention: UNESCO, 2001.

Jackson, Charles T. "Geological and Mineralogical Report." Washington DC: United States Senate, 1849.

Keweenaw National Historical Park, ed. "Final General Management Plan and Environmental Impact Statement." Calumet, MI: U.S. Department of the Interior, 1998.

Maher, Stuart W. "The Copper-Sulfuric Acid Industry in Tennessee," edited by Division of Geology, State of Tennessee Department of Conservation. Nashville, TN: State of Tennessee, 1966.

Marion, Daniel P. "Final Screening Site Inspection Report, Tennessee Chemical Company." edited by Superfund Division NUS Corporation. Copperhill, TN: U.S. EPA, Waste Management Division, 1989.

Martin, Dale, and Brian Shovers. "Butte, Montana: An Architectural and Historical Inventory of the National Landmark District." Butte, MT: Butte Historical Society, 1986.

Martin, Patrick. "Technical Report on the Archeological Survey of Isle Royal National Park," edited by Michigan Technological University. Houghton, MI: Isle Royale National Park, 1986.

Martin, Patrick, and Gianfranco Archimede. "The Quincy Mining Company Smelting Works, 1898: Historical Land Use Survey Project," edited by Michigan Technological University. Houghton, MI: Keweenaw National Historical Park, 2002.

Michigan State Historic Preservation Office, ed. "Programmatic Agreement among the Michigan State Historical Preservation Officer, Michigan Technological University and Keweenaw National Historical Park, National Park Service (U.S. Department of the Interior), Regarding the A.E. Seaman Mineralogical University Adaptive Reuse Project, Er06-107." Lansing, MI: MISHPO, 2006.

Muncy, Jack A. "A Plan for Revegetation Completion of Tennessee's Copper Basin," edited by Division of Land and Economic Resources. Norris, TN: Tennessee Valley Authority, 1986.

National Park Service, ed. "National Register of Historic Places Registration Form: Butte-Anaconda Historic District-National Historic Landmark Nomination." Butte, MT: U.S. Department of the Interior, 2005.

National Park Service Midwest Region, ed. "Options for National Park Service Involvement in the Management of Historic Copper Mining Resources on Michigan's Keweenaw Peninsula." Omaha, NE: U.S. Department of Interior, 1988.

———. "Study of Alternatives for a Proposed Keweenaw National Historical Park." Omaha, NE: National Park Service, 1991.

Natural Resource Damage Program, ed. "Summary of 2008 Settlement of Clark Fork River Remediation and Natural Resources Damages Claims and Related Restoration Plans." Helena, MT: Montana Department of Justice, 2008.

———. "Upper Clark Fork River Basin Restoration Plan Procedures and Criteria." Helena, MT: State of Montana, 2000.

Office of Enforcement and Compliance Assurance, ed. "Historic and Archaeological Resource Protection for USEPA Personnel." Washington, DC: U.S. EPA, 2007.

PTI Environmental, ed. "Old Works Engineering Evaluation/Cost Analysis-Preliminary Site Characterization." Bellevue, WA: ARCO, 1991.

"Quincy Mining Company Annual Report for 1923." Hancock, MI: Quincy Mining Company, 1923.

"Regional Historic Preservation Plan: Butte-Anaconda Corridor." Butte, MT: Regional Historic Preservation Plan Joint Committee, 1993.

Renewable Technologies Inc., ed. "The Butte-Anaconda Historical Park System Master Plan." Butte, MT: Butte Historical Society, 1985.

Smith, Ralph I. "History of the Early Reduction Plants of Butte, Montana." Butte, MT: State of Montana Bureau of Mines and Geology: Montana College of Mines, 1953.

U.S. Bureau of Reclamation, ed. "Old Works Operable Unit: Engineering Evaluation/Cost Analysis Work Plan." Billings, MT: U.S. EPA, 1988.

U.S. Department of Agriculture, ed. "Report of the Secretary of Agriculture for 1875." Washington, DC: U.S. D.A, 1875.

U.S. EPA, ed. "Copper Basin Mining District Case Study: Use of Cooperative Agreements Toward a Common Goal." Washington, DC: U.S. EPA, 2005.

U.S. EPA Region 4, ed. "Copper Basin Mining District Site Removal Action Update." Atlanta, GA: U.S. EPA, 2007.

———. "EPA Proposes Action on Davis Mill Creek.". Atlanta, GA: U.S. EPA, 2003.

———. "EPA Proposes Action on North Potato Creek." Atlanta, GA: U.S. EPA, 2003.

U.S. EPA Region 5, ed. "Declaration for the Record of Decision, Torch Lake Site, Michigan, Operable Units I and II." Chicago, IL: U.S. EPA, 1992.

———. "EPA Completes Construction at Torch Lake." U.S. EPA, 2006.

———. "EPA Proposes Cleanup Plan for Quincy Smelter Erosion." Chicago, IL: U.S. EPA, 2008.

———. "EPA Record of Decision: Torch Lake OUs 1 & 3 Houghton County, Michigan." Chicago, IL: U.S. EPA, 1992.

———. "First Five-Year Review Report for the Torch Lake Superfund Site." Chicago, IL: U.S. EPA, 2003.

———. "Proposed Plan for Record of Decision Amendment for Operable Unit 3 at Torch Lake Superfund Site." Chicago, IL: U.S. EPA, 2008.

———. "Record of Decision, Decision Summary, Torch Lake Sites, Operable Units I & II, Houghton County Michigan." Chicago, IL: U.S. EPA, 1992.

———. "Responsiveness Summary, Torch Lake Superfund Site, Operable Units I and II, Houghton County, Michigan." Chicago, IL: U.S. EPA, 1992.

———. "Second Five-Year Review Report for the Torch Lake Superfund Site." Chicago, IL: U.S. EPA, 2008.

U.S. EPA Region 8, ed. "Programmatic Agreement among the Environmental Protection Agency Region VIII Montana Office, the Advisory Council on Historic Preservation, the Montana State Historic Preservation Office, Montana Department of Health and Environmental Sciences, Local Governments of Butte/Silver Bow and Anaconda/Deer Lodge, Walkerville, and Arco Regarding Development of and Implementation of the Upper Clark Fork River Basin Regional Historic Preservation Plan." Butte, MT: U.S. EPA, 1992.

———. "Role of Clean Air Act Requirements in Anaconda Copper Company's Closure of Its Montana Smelter and Refinery." Denver, CO: U.S. EPA, 1981.

———. "Second Programmatic Agreement among the Environmental Protection Agency Region VII Montana Office, the Advisory Council on Historic Preservation, the Montana State Historic Preservation Office, Montana Department of Health and Environmental Sciences, Local Governments of Butte/Silver Bow and Anaconda/Deer Lodge, Walkerville, and Arco Regarding Development of and Implementation of the Upper Clark Fork River Basin Regional Historic Preservation Plan." Butte, MT: U.S. EPA 1994.

———. "Third Five-Year Review Report for Anaconda Company Smelter Site Anaconda, Deer Lodge County, Montana." Helena, MT: U.S. EPA, 2005.

Waggoner Barge, Sumner, & Cannon, Inc., ed. "A Future Use Plan for Redevelopment of the Abandoned Mine Lands in Copperhill Tennessee." Ducktown, TN: Glenn Springs Holdings, Inc., 2003.

———. "Report for the Extent of Lead Contamination at the Former Isabella Lead Chamber Acid Plant, Ducktown, Tennessee." Nashville, TN: Glenn Springs Holdings, Inc., 2003.

———. "Report of Inventory and Characterization of Former Mining and Processing Areas and Materials Located on Public and Private Lands." Ducktown, TN: Glenn Springs Holdings, Inc., 2004.

Theses and Dissertations

Bridge, Gavin. "Regulating Nature: The Ecological Origins and Environmental Implications of Restructuring in the U.S. Copper Industry." PhD diss., Clark University, 1997.

Curran, Mary Elisabeth. "The Contested Terrain of Butte, Montana: Social Landscapes of Risk and Resiliency." Master's thesis, University of Montana, 1996.

Curtis, Kent. "An Ecology of Industry: Mining and Nature in Western Montana." PhD diss., University of Kansas, 2001.

Durali-Muller, Soodabeh. "Roman Lead and Copper Mining in Germany, Their Origin and Development Throughout Time Deduced from Lead and Copper Isotope Provenance Studies." PhD diss., Johann Wolfgang Goethe University, 2005.

Jones, Michael Rice. "Oxhide Ingots, Copper Production, and the Mediterranean Trade in Copper and Other Materials in the Bronze Age." Master's thesis, Texas A&M, 2007.

LeCain, Timothy J. "Moving Mountains: Technology and the Environment in Western Copper Mining." PhD diss., University of Delaware, 1998.

Quivik, Fred. "Smoke and Tailings: An Environmental History of Copper Smelting Technologies in Montana 1880–1930." PhD diss., University of Pennsylvania, History and Sociology of Science, 1998.

Van Lingen, Gary W. "The Interpretation of Archaeological Remains at the Pittsburgh and Boston Copper Harbor Copper Mining Company's Second Campsite: Evaluating a Structural Component and Its Potential Uses and Contexts." Master's thesis, Michigan Technological University, 2002.

Unpublished Manuscripts

Bell, Peter, and Justin McCarthy. "Early Copper Smelting Technology in Australia." Paper presented at the Third International Mining History Conference. Golden, CO, June, 1994.

Borsos, E., L. Makra, R. Beczi, B. Vitanyi, and M. Szentpeteri. "Anthropogenic Air Pollution in Ancient Times." Paper presented at Universitatis Szegediensis, Szeged, Hungary: Universitatis Szegediensis, 2003.

Brignola, Alessandra "The Michigan Smelter at Cole Creek, Houghton: An Introduction." Houghton, MI: Michigan Technological University, 2006. Unpublished graduate paper available at Michigan Technological University Archives.

Corless, James. "The Significance and Status of the Quincy Smelting Works, Franklin Township, Michigan (White Paper)," edited by Keweenaw National Historical Park. Calumet, MI: Keweenaw National Historical Park, 2008. Available at Keweenaw National Park.

"Dates in Basin History." Ducktown, TN: Ducktown Basin Museum, 2000 (est.). Available at Ducktown Basin museum.

"Dates in TCC and Copper Basin History." Copperhill, TN: Tennessee Chemical Company, 1987. Available at Ducktown Basin Museum.

"Letter from EPA Region 5 to Frank Fiala, National Park Service-U.S. Department of the Interior (June 23, 2004)." Available at Keweenaw National Park.

"Letter from EPA Region 5 to Mr. Jim Corless, Superintendant, Keweenaw National Park (January 20, 2009)." Available at Keweenaw National Park.

Nordberg, Erik C. "Quincy Mine Hoist Association Anniversary Banquet." Hancock, MI: Quincy Mine Hoist Association, 2008. Unpublished booklet available at Michigan Tech Archives.

Rush, Ken. "Copper Mining Companies of the Basin." Ducktown, Tenn.: Ducktown Basin Museum, 1993. Available at Ducktown Basin Museum.

Strong, Katherine. "Shoreline Change Along the Portage Canal." Houghton, MI: Michigan Technological University, 2002. Unpublished graduate paper available at Michigan Technological University Archives.

Trepal, Dan. "The Calumet & Hecla Smelting Works: 1887–2006." Houghton, MI: Michigan Technological University, 2006. Unpublished graduate paper available at Michigan Technological University Archives.

"Copperama," 1960 (est.). Unpublished pamphlet available at Michigan Tech Archives

Wilson, C. P. "A Brief History of Copper Smelting on Portage Lake." Houghton, MI: Michigan Technological University, 2008. Unpublished graduate paper available at Michigan Technological University Archives.

Web Pages

"A barren landscape was created after miners cut down forests for fuel. Location: Near Copperhill, Tennessee, (Picture Id: 620678.)" *National Geographic Stock*, 2008 (est.), www.nationalgeographicstock.com/ngsimages/explore (accessed December 10 2008).

"Butte Silver Bow Public Archives." *City and County of Butte-Silver Bow*, 2009, www.buttearchives.org (accessed May 2, 2009).

"CERCLA and NHPA Coordination for Superfund Sites." *National Preservation Institute*, 2009, www.npi.org/sem-cercla.html (accessed July 11, 2009).

"CITGO Company History." *CITGO Petroleum Company*, 2009, www.citgo.com/aboutcitgo/companyhistory.jpg (accessed March 4, 2009).

"Cleanup Process." *Superfund-U.S. EPA*, June 3, 2009, www.epa.gov/superfund/cleanup/index.htm, (accessed July 11, 2009).

"Copper Basin." Glenn Springs-Copper Basin Project, www.glennsprings-copperbasin-project.com/history.htm (accessed March 7, 2009).

"Copper Basin Mining District." *U.S. EPA Region 4*, February 18, 2009, www.epa.gov/region4/waste/copper (accessed March 13, 2009).

"Copper Basin Reclamation Project." *Tennessee Valley Authority*. 2003 (est.), www.tva.gov/environment/land/copper.htm (accessed March 11, 2009).

"Copper Statistics." *U.S. Geological Survey Historical Statistics for Mineral and Material Commodities in the United States*, 2006, http://minerals.usgs.gov/ds/2005/140/copper/pdf (accessed March 13, 2008).

"Copper Statistics and Information." *USGS Minerals Information*, June 15, 2009, http://minerals.usgs.gov/minerals/pubs/commodity/copper (accessed July 8, 2009).

"County Population Census Counts 1900–90." *U.S. Census Bureau*, July 15, 2008, www.census.gov/population/www/censusdata/cencounts/index.html (accessed April 4, 2009).

"Gold Statistics." *U.S. Geological Survey Historical Statistics for Mineral and Material Commodities in the United States*, 2006, http://minerals.usgs.gov/ds/2005/140/gold/pdf (accessed 15 March, 2008).

"History of the Museum." *World Museum of Mining*, February 14, 2008, www.miningmuseum.org/historyofwmm.html (accessed April 2, 2009).

"Keweenaw Heritage Sites." *Keweenaw National Historical Park*, April 3, 2008, www.nps.gov/kewe/parkmgmt/keweenaw-heritage-sites.htm (accessed June 25, 2009).

"Material Property Data." *MatWeb*, 2008, www.matweb.com (accessed March 13, 2008).

"National Register of Historic Places." *U.S. National Park Service*, www.nps.gov/history/nr/ (accessed March 18, 2011).

"Old Works/East Anaconda Smelter Case Study." *Superfund-U.S. EPA*. September 19, 2007, www.epa.gov/superfund/programs/recycle_old/success/ casestud/anaccsi.htm (accessed September 20, 2008).

"Photograph and Image Search." *Tennessee State Library and Archives*, 2009, http://tnsos.org/tsla/imagesearch/index.php (accessed March 31, 2009).

"Pig Iron Statistics." *U.S. Geological Survey Historical Statistics for Mineral and Material Commodities in the United States*, 2007, http://minerals.usgs.gov/ds/2005/140/ironsteel/pdf (accessed March 13, 2008).

"Silver Statistics." *U.S. Geological Survey Historical Statistics for Mineral and Material Commodities in the United States*, 2006, http://minerals.usgs.gov/ds/2005/140/silver/pdf (accessed 15 March, 2008).

"Southeast Tennessee Development District." *Southeast Development Group*, 2009, www.sedev.org (accessed March 13, 2009).

"Superfund Program, Anaconda Co. Smelter Site." *Superfund-U.S. EPA*. 2009 (est.) www.epa.gov/unix0008/superfund/mt/anaconda/index.html (accessed July 8, 2009).

"Superfund Program: Anaconda Co. Smelter Site." *U.S. EPA Region 8*, March 17, 2009, www.epa.gov/region8/superfund/mt/anaconda/index.html (accessed April 18, 2009).

"Superfund Program: Silver Bow Creek/Butte Area." *U.S. EPA Region 8*, March 17, 2009, www.epa.gov/region8/superfund/mt/sbcbutte/index.html (accessed April 18, 2009).

"Tennessee Historical Commission State Owned Sites." *Tennessee Department of Environment and Conservation*, February 7, 2005, www.state.tn.us/environmental/hist/stateown/ducktown.shtml (accessed March 12, 2009).

"USA State & County Quickfacts." *U.S. Census Bureau*, February 20, 2009, http://quickfacts.census.gov (accessed March 24, 2009).

"Welcome to Anaconda." *Community of Anaconda, Montana*, 2008, www.anacondamt.org (accessed April 2, 2009).

"Who are we and how are we organized?" *U.S. EPA*, 2009, www.epa.gov/epahome/aboutepa.htm (accessed July 11, 2009).

Index